"Before we invented civilization our ancestors lived mainly in the open out under the sky. Before we devised artificial lights and atmospheric pollution and modern forms of nocturnal entertainment we watched the stars. There were practical calendar reasons of course but there was more to it than that. Even today the most jaded city dweller can be unexpectedly moved upon encountering a clear night sky studded with thousands of twinkling stars. When it happens to me after all these years it still takes my breath away."

Carl Sagan, *Pale Blue Dot*

The
CONSTELLATIONS
Volume I

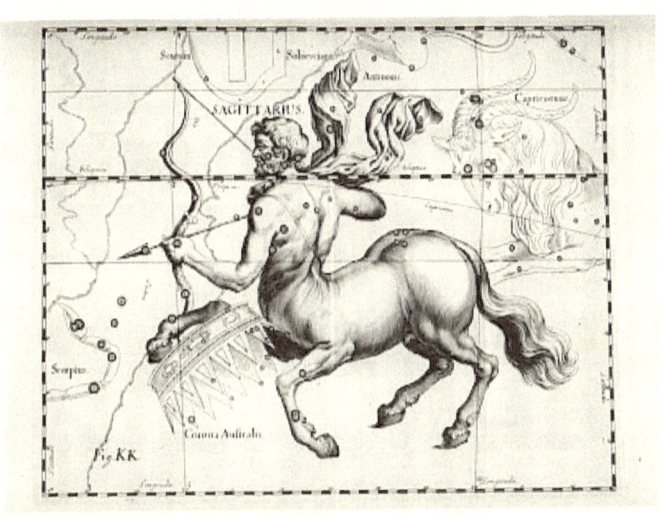

Sky Tours For Computerized Telescopes

Enjoying the sights, history and
lore of the night sky

by

P. Clay Sherrod

Printed by LuLu – Publishing and eBook Company
Copyright 2017,   P. Clay Sherrod
Printed in the United States of America
All rights reserved, Published 2017

Arkansas Sky Observatories Publications

ISBN: 978-1-365-71169-5

Constellations

is dedicated to
my Grandparents
who taught me that it is okay
to use my imagination

so that I might see creatures in the sky.

"Those who first invented and then named the constellations were storytellers. Tracing an imaginary line between a cluster of stars gave them an image and an identity. The stars threaded on that line were like events threaded on a narrative. Imagining the constellations did not of course change the stars, nor did it change the black emptiness that surrounds them. What it changed was the way people read the night sky...."

- John Berger -

INTRODUCTION

As a child we yearn to understand the stars and the patterns that they inscribe throughout the sky. We see visions of lions, great hunters and flying eagles throughout the night, all emblazoned in our memories from the spectacle of a starry night.

Such it has been since mankind first walked upon the Earth, attempting to give some organization, some humanly engineered aspect to God's creation of the Heavens.

In this series I have combined this prehistoric drive for understanding with the age of modern, computer driven telescopes, capable of their own power and pointing at the touch of a button. Here we can learn the science, mythology and charms that await.

These Guides will hopefully provide you with at least a teaser of the beauty and fascination that the constellations hold.

This two-book series provides the major constellations as seen from the northern hemisphere and not all constellations are included south of the celestial equator.

Please make this YOUR work in progress and add to the many spectacular and fascinating objects that our constellations provide. Let this be a start to your astronomical database for your modern computerized telescope.

Doc Clay - 2017

CONTENTS
VOLUME ONE

Introduction

Computerized Sky Programs – Page 9
Basic Use of a Computerized Telescope – Page 19

Andromeda – page 31
Aquarius – page 59
Aquila – page 81
Aries – page 103
Auriga – page 123
Bootes – page 151
Camelopardalis –page 171
Capricorn – page 195
Cassiopeia – page 221
Cepheus – page 347
Cetus – page 251
Cygnus – page 301
Coma Berenices – pages 151/277
Corona Borealis – page 151
Delphinus – pages 301/325
Draco – page 347
Equuleus – page 325
Scutum – page 81
Ursa Minor – page 347
Vulpecula – page 301

Author Note: The Constellations guides are presented in alphabetic order for convenience of reference. However, note that many lesser constellations are included in MAJOR constellation discussions since they are bordering or very near these larger areas; hence each volume may have listings for constellations but data is not duplicated

*   *   *

Discussion

COMPUTERIZED SKY PROGRAMS

Although the subject material of this Constellation Series will remain intact for many years to come, the tools which are discussed here will likely change by the time this is even in print.

Technology has "taken over" virtually every aspect of our lives and the very act of communication is beginning to dominate our lives through the Internet, through "social media" and in ways we never dreamed of in the twentieth century, the era in which computerized telescopes emerged.

As computers became more and more dominant in the sciences and in astronomy in particular, I argued that such devices – in 1985 – had "no place in the gentleman's science of astronomy" – one of the pure sciences of the renaissance thinkers.

I fought as long as I could, but as of today, one of my private observatories has eleven – yes, ELEVEN – computers that are essential for my nightly research at the facilities. Others have no less than three each. I depend on them every night.

The computers operate the telescope, the domes, the cameras, the focusing, the temperature regulation, the star catalogs and research calculations. They essentially do everything. In fact, there are many remote robotic observatories worldwide that never even see a human darken their doors.

The opening and closing and the operations in between of each night's observing routine is carefully programmed and the computers do it all.

As I have grown to appreciate the abilities and efficiencies of the computer in astronomy, my productivity has increased along with technology. A mathematical reduction of a single comet's orbit or position in space might have taken a full night and subsequent day in 1973.....today I can do all of that for no less than 200 comets (or asteroids) in a single night, all thanks to computer operation and data management.

I am a Renaissance Astronomer who now simply loves computers.

At some time in the late 1960's I remember reading a then-current news story in the astronomy magazine Sky & Telescope about the first Alt-Azimuth computer-driven telescope in New Mexico that was going to scan dozens of galaxies in just one night for supernovae. Preposterous, I laughed....no telescope can do that.

Then, technology began to lurch forward and major public observatories worldwide became computerized with the ability to "Go To" objects at the computer's command. By the late 1990's Meade Instruments Company - a California-based manufacturer of high quality telescopes for the amateur astronomer – introduced the revolutionary LX 200 Schmidt-Cassegrain GO TO telescope, a feat of engineering that allowed the average star-gazer to have at his or her fingertips access to tens of thousands of celestial objects at the push of a button. The era of modern GO TO telescopes for the average guy had started.

Today, virtually every telescope manufacturer sells GO TO telescopes and all work quite well.

A keypad can instantly let the telescope find a star, a planet, a distant galaxy or comet. No more hunting around for faint fuzzies in the eyepiece, when the on-board computer can do all the work for you. Set up the telescope, put it in its required home position, make sure you have the right date and time and location.....and you are instantly finding sky objects that you would never have seen.

A casual amateur astronomer now has in one handbox the ability to see without any fuss or frustration, far more objects in one night than we could in a month only thirty years ago.

All thanks to those computers, those things that did not "belong" in a Renaissance science.

COMPUTERIZED TELESCOPES

I very much hesitate to start listing manufacturers of computerized GO TO telescopes....the list has new entries every year, and many drop out in the course of only a few short years. Nonetheless, I will acknowledge from my personal experience those who provide telescope systems with on-board computers for GO TO operation – those which I feel have the quality and reputation to support:

Astro-Physics GTO – absolutely the finest telescopes and mounts available; very high end, but clearly worth every cent.

Celestron Nexstar – making fine telescopes for many decades and the pioneer in compact, portable instruments. Their NexStar GO TO system is fine and quite user-friendly although in my opinion not so much as the Meade product.

iOptron – primarily a supplier of moderately priced computerized telescope mounts (not complete systems), this company offers a very nice series of GO TO mounts with computer-adaptable keypad that is similar in many respects to that of the Celestron telescopes, although somewhat difficult in terms of viewing of display.

Orion Telescopes – around for many years providing privately-branded production telescopes, Orion has a large selection of reflectors, refractors and catadioptic designs, all with very nice GO TO mounts with keypad virtually identical to that of the Celestron Nexstar, with a database of nearly 43,000 objects.

Meade Instruments – in my opinion by far the most user-friendly of all computerized controllers, the Meade Autostar/Audiostar allows for personal input of user object data and has a database of more than 37,000 objects (Audiostar) to over 300,000 objects (Autostar II for advanced telescopes.

Planewave Instruments – a high-end maker of fine RC-type telescopes, utilizing several larger computerized mounts (buyer's choice) to include:

Parallax, Mathis Instruments, Micron, Astro-Physics and others. All fully computerized, the operation will depend on the mount selected.

Vixen – a Japanese manufacturer of excellent quality, they offer a full line of smaller – but somewhat expensive – GO TO telescopes, utilizing the Sphinx mount and Starbook computer control technology.

Many other manufacturers of mounts with computer capability can be found by doing a computer search on this subject. However, I urge anyone considering a new telescope to attend a local star party and examine first hand the array of possibilities in these telescopes and talk to users about the pros and cons of each.

SOME PLANETARIUM / SKY PROGRAMS

Just like telescopes, there is an increasingly innovative list of suppliers of excellent computer control software and planetarium sky programs for telescope operation. When choosing, make sure that you get one that is very user-friendly, one that can allow input of TOURS and USER OBJECTS that you will want to personally archive in a sky library, and one that will function with your particular telescope.

The following examples, are my first choices, but there are many others which I unfortunately have not had the opportunity to test and assess. And – just like computers – new programs are always being introduced, and old standards are always being updated and improved!

Cartes du Ciel – this is a very popular FREE telescope control software and planetarium program which has far-reaching advanced applications, and the ability to incorporate current star catalogs such as UCAC4 and others. You cannot beat "free."

Deep Sky Planner by Knight Sky – used by many and has complete and accurate telescope control. Perhaps not as adaptable as some others, but a very good program.

GUIDE by Project Pluto – by far my preferred telescope and planetarium program for telescope operation. Intended mainly for advanced observers, this program is still suitable for all; the nice advantage of GUIDE is its ability to be updated very easily with User Objects, comets, asteroids, and new objects direct from links available on-line.

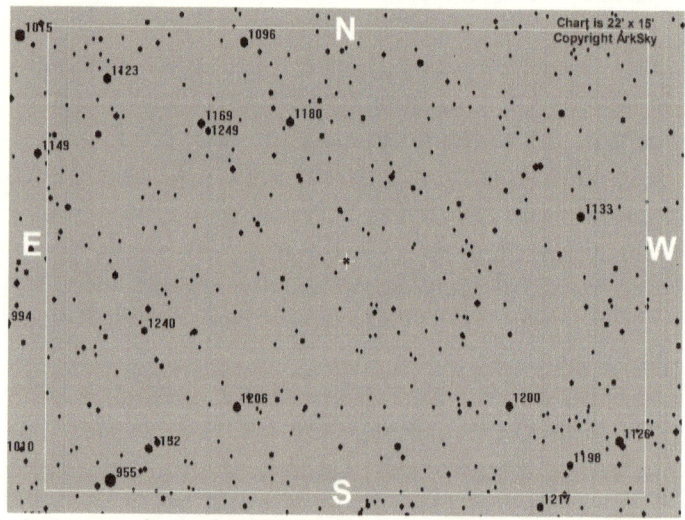

A screen shot of GUIDE display from ASO; in this example the background is grey with black stars

One very fine feature of GUIDE is its ability to have the extensive toolbar which allows your selection at a fingertip of color combinations for display, limiting magnitudes, GO TO objects and much more. No cute horizons, no constellation drawings....just incredible ease of use.

Starry Night – outside of research use at my observatory, my preference for teaching and sky demonstrations is Starry Night software, allowing full telescope control and easy input of current objects such as asteroids and comets. Many versions of this excellent program are available, from beginner/novice to Starry Night Pro for advanced observers. Beautiful realistic horizons an optional mythological constellation depictions.

Stellarium – this is a very popular and sometimes free software that allows both planetarium use and telescope control.

The Sky – this is a mainstay resource for many advanced observers, and this is required software for the Software Bisque line of computerized mounts. This is an excellent all-inclusive (allows astrometric and photometric measures and other applications), but is many times not as user friendly as other software packages.

APPS FOR SMART DEVICES

We are in a new generation of technology, pretty much ruled by social media and "smart devices" such as the iPhone, tablets, and pads. Nearly all of these can produce effective results in terms of telescope operation and display of any computer.

These devices can be connected to telescopes via the many creative APPS that are available, either wired or wirelessly through Bluetooth-type technology, for very efficient telescope control, which makes these devices so perfect for use in the field, or at star parties.

A good iPhone App can provide just as efficient telescope control and object information database as many PC planetarium programs. For advanced use and complete automation, they are not recommended.

Follow is a brief list of some of what I consider to be the most applicable and complete Apps available today which provide both sky charts with database and complete telescope control. Note that in all cases, some connective device is necessary for your telescope, whether a wireless bud or proper wired connection which would be obtained separately from the program.

You MUST obtain the App that pertains only to your operating system for your device as noted.

For iPhone/Apple Devices:

Sky Safari – by far the most advanced and my favorite of all smart device Apps; as with many PC programs, this one comes in several versions, but all have complete telescope control and fantastic database reference and sky maps.

Starmap – also with several versions; this one has the capability of ASCOM output for accessory control.

Luminous – honestly, this one has at present a limited user base, but is one of the highest rated sky Apps with complete telescope control among those who have tried it.

Orion's StarSeek – also among the highest rated Apps for iPhone, this is distributed under private label by the major telescope manufacturer.

For Android, etc.:

Sky Safari for Android - by far the most advanced and my favorite of all smart device Apps; as with many PC programs, this one comes in several versions, but all have complete telescope control and fantastic database reference and sky maps.

Virtuoso – a very fine and highly rated Android App that allows excellent sky charts, information and total telescope control.

Sky Portal – efficient and very clear, user-friendly; ratings are not as high on this one as they are for Sky Safari.

*   *   *

All of these smart device Apps provide very detailed depictions of the constellations, and can be used with beautiful mythological graphic representations of the 88 constellations, or those can be turned off to show the sky in its natural appearance

Meade Instruments' Highly Portable
Yet lightweight computerized GO TO telescope
The LX90 – *Courtesy Meade Instruments*

BASIC USE OF A
COMPUTERIZD TELESCOPE

Each of the Constellation Guides will be fundamentally the same, giving you quick access to the best and sometimes the most challenging of objects in each star pattern in the sky.

Every brand of telescope is different....every sky program for computers is different than the other, but in essence, all work the same way and the process for commanding GO TO operation is the same between all of them.

This Guide will assist you in learning the joy of using a computerized telescope, but also help you in developing your own customized lists of objects that might be of interest to you: variable stars, double stars, galaxy novae searches, near Earth objects....the list is as endless as space itself.

In my opinion, the logic of Meade's Autostar GO TO system is the most user-friendly and logical that is on the market; hence there will be references to this platform throughout these books. However, all of the computerized programs and handboxes work essentially the same, so learn how yours might be different from what is described and apply this logic to all discussions that follow.

As with all of the "GO TO" TOUR constellation tours, I recommend a hard copy OR access to a computer monitor showoing a good star atlas and/or chart which will list all the finest objects, constellation-by-constellation. One very handy reference guide to consult is the PETERSON

FIELD GUIDE TO THE STARS AND PLANETS (by Donald Menzel), which features complete lists with declinations, right ascensions, magnitudes, and all pertinent information for you to expand your observing horizons beyond this brief Guide.

As mentioned, there are three platforms which will provide ease, fun and excitement at the telescope through which you can operate through computerized commands:

KEYPAD – these are computer controlled internal systems built for each model of telescope or mount.

PC SKY PROGRAM – installed planetarium programs that will display on your PC or laptop computers for use at the telescope, all of which have telescope commands. These powerful platforms allow totally remote robotic operation of a properly equipped telescope system.

BLUETOOTH - Ideal for star parties and personal nights out with the telescope, these provide telescope control (if the telescope is so equipped) via your smart phone, tablet or similar device, even without the need to connect.

See previous discussion on all of these formats.

KEYPAD OPERATION

Nearly all popular telescope manufacturers today offer some form of GO TO operation via a wired (or wireless) keypad that communicates directly to the telescope's mounting. Many are operated in both Alt Az and Polar mount configurations.

In computerized telescopes, the "smarts" of the computer processors are powerful and can compute the daily motions of comets and planets. The actual computer processor for such telescopes may be located in the actual keypad (handbox) or inside the mount itself.

Nearly all work the same, except for some perhaps confusing jargon in labeling, such as Meade's Autostar "Mode" to back out of a particular screen or command comparerd to Celestron's Nexstar "Back". Astro-Physics uses the word "Menu" to return, or back out of an existing command.

Yet, all do the same thing. So it is up to the user to define the exact operation of his or her GO TO telescope keypad and become proficient at its operation. Here, I am discussion the basics of keypad entry for GO TO commands.

***NOTE: The following instructions are for the MEADE Autostar computerized GO TO System.
Please NOTE that nearly all other PC planetarium programs and telescope internal GO TO functions operate similarly. Refer to the precise instructions for those programs for your specific telescope or PC program! You will find that the instructions are quickly adaptable for your system....*

FOR AUTOSTAR: Note that your AutoStar will NOT have every object listed on every constellation GO TO tour....this is intentional. You can access some of the most interesting objects of the sky directly from their coordinates. It is quite simple as you merely enter these coordinates as follows in the 10-step process:

1) Press the "MODE" key and hold down for 3 seconds and release;

2) Displayed will be the current Right Ascension and Declination of the center of field of view of where your telescope is presently pointed (assuming

that you have properly aligned from "home position");

3) [NOTE: if you have the Meade electric focuser attached to any of the ETX or LX telescopes, holding down the "MODE" key will bring up the "Focus" command first....merely scroll (lower right scroll key) down one step to access the RA and DEC to enter your desired coordinates]

4) Press the "GO TO" button on AutoStar;

5) This will change the display and you will note the cursor blinking over the first digit of RIGHT ASCENSION (R.A.); merely use the number keys and dial in the R.A. of the object you are searching for;

6) When done, press "Enter;"

7) This moves the blinking cursor over the "DEC" coordinates;

8) [NOTE: the declination, unlike R.A., can be either positive or negative and you will see the "+" or "-" sign displayed depending on where your telescope is aimed at that time; if it is NOT the desired setting (plus or minus), merely use your arrow key to move the blinking cursor OVER the "+" or "-" sign and change by using either of your lower corner SCROLL KEYS;

9) Proceed to enter the DEC using number keys;

10) Press either "Enter" or "Go To" when finished and the telescope begins slewing to your desired object!!

PC OR LAPTOP SKY PROGRAMS

There is considerably more "power" in utilizing one of the major sky programs ("planetarium programs") that are available today on a personal computer or laptop. The latter platform is advised for field use and casual personal use at the telescope, but a powerful desktop computer can be a command center for a sophisticated amateur observatory, allowing telescope and dome operation, camera control and image acquisition, all done from right next to the telescope, or from any location on Earth.

Note that each sky program is uniquely different in terms of the way their commands appear (i.e., the Toolbar, etc.). Nonetheless, each operates pretty much the same way, and nearly all will operate any computerized telescopes equipped for GO TO operation. In such cases, even the keypad is not even needed once the computer is connected to the system.

All planetarium programs have a basic sky chart that can be enlarged to match the precise field of view of your telescope; colors can be changed, with a pleasing red screen, and objects showing in white or grey against that background. You can change the color of stars from white to black if you want for eye-saving comfort in nighttime viewing.

All sky programs allow telescope interface. In some cases the huge free telescope program ASCOM may be needed to couple your telescope with your sky program operating system. If possible I strongly recommend avoiding ASCOM if at all possible, because the program many times can become difficult when multitasking. Nonetheless, if you are operating a dome, telescope, camera, focuser and more, one of those devices may require ASCOM for operation.

Telescope commands are typically given via a "flyout" telescope box that gives motion control and GO TO commands once your object(s) is selected. In order for your sky program to function properly a couple of key considerations must be met:

TIME – your date and time must be set on both the telescope and the computer and both must match and be synchronized.

LOCATION – your PC program must know where you are, and there are provision on all programs to enter your telescope's geographic latitude and longitude.

INITIALIZATION – your telescope will need to be properly initialized at least the first time per the instruction that come with the telescope. Once this initialization is done (just like you would do in the field), you may wish to PARK your telescope at the end of an observing session (if a permanent installation) which will allow both the telescope and the sky program to "remember" your last observing details and the position of the telescope even when you power off. If PARKing your telescope, this allows for instant access between computer and scope, with no need to again initialize nor align the telescope for any permanently mounted instrument.

SYNCHRONIZATION – typically you will not want to use the handbox while the sky program is operating you telescope on your PC. The sky program must recognize the telescope's connection and you need to verify that the connection exists before attempting GO TO operation. Never – ever – go off and leave your telescope when you first connect and attempt your first GO TO via remote computer. Be there to assure that the telescope and computer are talking the same language. Run at least two bright star GO TO's first before assuming that the system is synchronized.

The options and capabilities of good planetarium programs on your computer are seemingly endless and this is by far the most powerful, versatile, and enjoyable way to fully appreciate the power of your instrument.

SMART DEVICES AND BLUETOOTH

The "Smart Age" of devices has caught up with the gentleman's science of astronomy. For those with iPhones, any smart phones, tablets, or pads, your telescope can be completely operated remotely and wirelessly with any of these smart devices!

There are many companies than provide wireless "buds" that will connect in place of your computerized telescope handbox, through which communication to your smart device can easily take place. There are several innovative and powerful sky Apps available (as discussed previously) that not only provide you with complete telescope control, but also a complete library of information, data, sky coordinates, and even the latest photographs of each and every object that you turn your telescope to, all from the touch screen of your phone or tablet.

If you decide to control your telescope via such a Bluetooth device, you will want to consult (or computer search) your exact device that you have and obtain specifically compatible wireless devices that will need to be purchased for your smart device manufacture and platform.

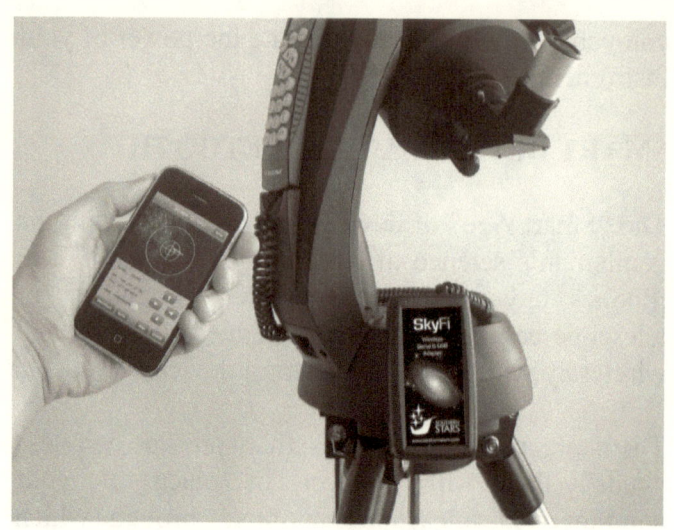

The innovative wireless SkySafari5 operating a
computerized Celestron telescope via smart phone
Image courtesy SkyFi

*   *   *

ENJOYING THE CONSTELLATIONS

The constellation tour Star Charts provided here
will get you started on your journey for this
constellation.

Each Tour will contain a concise object list for your
"GO TO" tour of each constellation; you may wish
to find the majority of the objects from the AutoStar
Library (for example, you can easily go to the
wonderful Andromeda Galaxy, *Messier 31*, if you
go to your system's menu, then pull up "Object /
Deep Sky / Messier /..then type in '31'...." and then
press "Enter", followed by "GO TO" to access this

naked-eye galaxy. (your procedure will vary depending on the sky program being used)

On the other hand, if you want to experiment and become a "better sky computer user" try entering the exact R.A. and DEC coordinates of that object as described for your program. In Autostar, after holding down the MODE key you will find the menu for this manual entry. PC sky programs have capability for keyed-in coordinate entry, usually by the command GO TO / Coordinates ' [enter data as formatted] and then actually saving that data to a User File.

You will find the accuracy of entered GO TO's to be somewhat better than those stored in the internal computers of telescopes (since the epoch is likely more current) but the capability of acquiring unlisted objects is fantastic!

In each *Constellation* chapter, you will access your FIRST GOTO target - usually the brightest star in each constellation - via the command specific to locating most of the bright NAMED stars for each constellation. Each sky program and telescope keypad has a Star Name list and access is quick and easy.

PC Sky Programs and your smart devices also have library access to allow search and selection of the brightest (Alpha) stars of each constellation.

As an example, in Meade's Autostar, we scroll to the primary menu: For our first upcoming constellation, Andromeda, we would need to locate and center by commands the bright star

Alpheratz.by selecting: "SETUP / OBJECT / STAR / NAMED....and scroll to "**Alpheratz**"", then press "Enter" and subsequently press "GO TO" to move your telescope to this bright star.

Remember also that – in this constellation - the *Andromeda Galaxy* is listed among the "named" objects. This galaxy can be accessed easily in any sky programs via using its NAME (Named Objects), as a GALAXY (under Deep Sky/Galaxies), its Messier Catalog listing as well as its NGC number ((New General Catalog).   So, likewise for that object you might merely go to SETUP/OBJECT/DEEP SKY/NAMED....and then scroll alphabetically to the "A's" if you are not already there; press "enter" and then GO TO and your scope is off and running!

You will find that – in addition to looking at some of the most challenging and spectacular sights of the sky – you are going to thoroughly enjoy the "techno" end of the computerized telescope's capability.

In your backyard, home observatory, camping, star party, or sidewalk viewing for your neighbors, the entire sky is now literally at your fingertips!

*   *   *

Chapter One

ANDROMEDA

...yes, there IS a constellation out there surrounding that famous galaxy!

This is the first of your Constellation Guides, "GO TO ANDROMEDA" of the series for all telescope users". Being the first of our studies, this constellation features MANY wonderful deep sky objects, interesting stars and much wonderful historical interest. Yet Andromeda is a notably "ignored" constellation by the amateur astronomer, mainly because of all the focus of attention to the great and famous galaxy within it Messier 31 - the "*Andromeda Galaxy.*"

Andromeda (pronounced "an-DRAW-me-duh" In Greek legend is a very important mythological figure, the daughter of both Queen *Cassiopeia* and King *Cepheus*. As mentioned in previous constellation guides, the whole stellar affair of Cassiopeia with other immortals and mortals is something right out of a soap opera and no doubt much less befitting a Queen, particularly one immortalized in the beautiful heavens. Indeed, unfaithfulness to Cepheus in an illicit affair of Cassiopeia with the strong and dashing Perseus ultimately led to her to be banished forever to the sky chained to her thrown, encircling the pole star forever.

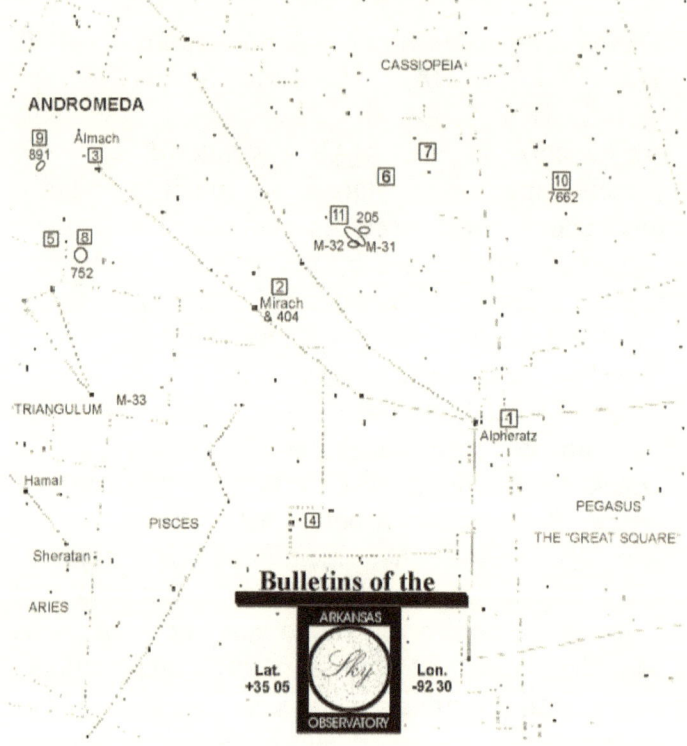

A finder chart for locating many of the GO TO objects in the constellation of Andromeda; if using a computer planetarium program, you are encouraged to plot the objects on your screen for higher resolution than this chart provides.

* * *

Andromeda, fortunately born before her mother's en-throned demise, has not been much more fortunate in legend....thanks once again to momma.

Cassiopeia - a braggart as well as a hussy - confronted the great sea nymphs of *Neptune*, beautiful to all who saw them, that she was far more beautiful than they and, indeed, more beautiful than ANY living (or immortal!) woman. Jealousy raged and the nymphs cried to Neptune to do something to place their beauty back to the highest ranks among mortal eyes.

As God of the Sea, mighty Neptune ordered the great whale monster *Cetus* to ravage the kingdom and the family of Cepheus to avenge the nymphs' forsaken achievement of such fabulous beauty. But Cepheus had wise sages (or a television psychic hot line) who informed him of the impending attack and learned that he COULD prevent the attack by the great sea beast if he merely sacrificed his daughter, Andromeda....NOT the lovely Cassiopeia mind you, but his innocent daughter. Well, that was okay by Cepheus so he arranged to have her taken and bodily chained to a large cragged rock overlooking the ocean to be devoured by the great whale.
That's what Dads are for.

Perseus - here is where it really gets interesting in light of the fact that we already know that Perseus has been "fooling around" with Andromeda's mother - heard of her plight and went boldly to her rescue, with the head of the "snaky" *Medusa* in his hand which, upon seeing it the monster turned instantly to stone. As just reward, mythology tells us that Perseus in turn asked Cepheus (actually "demanded") for Andromeda's hand in marriage, to which the great and wise King hastily agreed. Now THAT is one big, happy, Greek-god-of-a-family even in today's time!

This whole story begs the questions: 1) what was Perseus doing with Cassiopeia while he was pursuing Andromeda's hand in marriage? 2) did the Authorities come and get King Cepheus (see figure below) for child abuse and endangerment? and 2) do all sea monsters sink in the ocean when turned to stone or do some of them rise into the sky as stars as did Cetus?

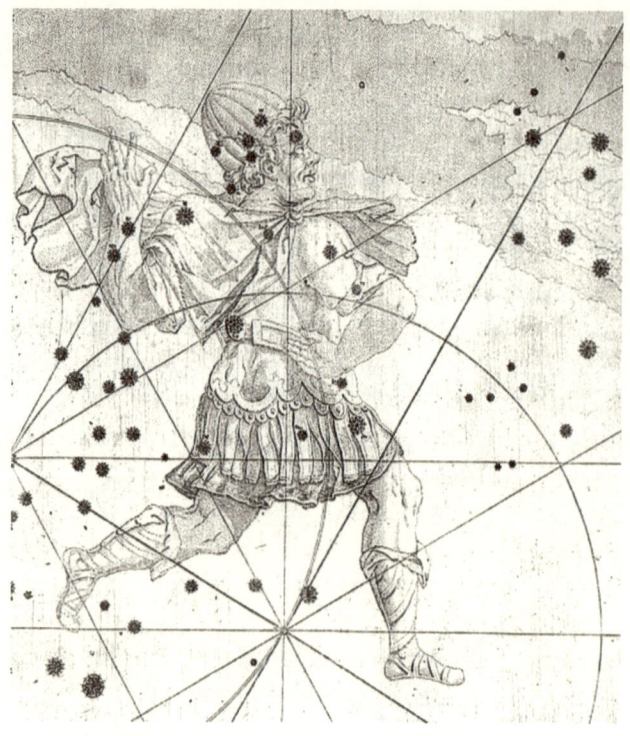

KING CEPHEUS Guilty of Child Abuse and
Endangerment
or merely looking out for subjects of his kingdom?
(from the original Bayer's 1603 Uranometria)
*   *   *

In addition to the famed Andromeda galaxy which is discussed in full here, Andromeda is loaded with nice objects for your telescope; there are at least three bright galaxies (all huddled around Messier 31), two nice galactic clusters and many very nice multiple and variable stars that we will explore.

To the naked eye, the constellation of Andromeda might appear at first a bit uninteresting, as there are no very bright stars nor conspicuous asterisms (star patterns) on which to focus your attention; nonetheless, the naked eye splendor of Messier 31 as it stretches across the central portion of this constellation like a ghost in flight is enough to clearly motivate even the novice observer to explore her boundaries. Adding even more to the naked eye spectacle on the clearest and darkest night is the even-larger (but fainter) Messier 33, the Triangulum galaxy, or "*Pinwheel*" to the immediate southeast of the great galaxy. Both are absolutely breathtaking in good binoculars on those crisp fall evenings of very deep sky.

The photograph following shows the constellation and its famous galaxy rising above 3,000 foot Wilhelmina Mountain in western Arkansas in 1972 through a piggyback guided Nikon camera with 50mm lens. Note the darkness of the skies all the way to the horizon over 40 years ago and the "sky glow" toward the middle of the photograph....that glow is caused by reflection of ionized dust particles that follow in the wake of the Earth in its orbit....today it is caused more likely by high pressure lighting on Earth. Facing this same direction today from this old observatory site is now a major shopping center with 24-hour lighting.

THE SKIES OF ANDROMEDA
The Arkansas Sky Observatory, Mena, Arkansas
3,000 feet Elevation - August 1972

As with every "GO TO" TOUR guide, each GO TO object in Andromeda is discussed for your telescope regarding the type of conditions necessary for you to view it optimally for discern the very faintest details.........magnifications and aperture necessary for most objects, and much, much more. This is YOUR complete guide to get you on your way to exploring the best (and few!) objects in these two constellations. The following listing of "BEST" objects contains the finest or most interesting from my own observing experience and preference.

Use the star chart and the following Guide as an excellent reference for your next star party itinerary, or a beginning for further study into the thousands of objects visible in this part of the sky. Truly these extensive Constellation Study Guides will most definitely put your AutoStar or PC sky program to work for you in the most efficient and enjoyable way possible! As a matter of fact, MANY sky program users are now programming their own "Tours" based on these guides, using each constellation as a separate GO TO Tour for the library that can be added in or deleted through the main edit screen on your PC or MAC computer.

We hope you enjoy these comprehensive GUIDES to touring the constellations via your computer-driven telescope. Each new installment is complete with diagrams, charts and illustrations that you will find nowhere else. Please let us hear YOUR feedback and your observations of each and every constellation after YOU have toured its vast reaches of our skies!

YOUR ANDROMEDA CONCISE DIRECTORY OF INTERESTING OBJECTS –

Attempt to pry your telescope away from the coordinates [R.A. 00h 40m / DEC +41 degrees 00m] just long enough to explore the many other wonderful celestial objects this rich constellation has to offer. Afterwards, you can return to the splendor of the closest galaxy that resembles our own to the Earth!

I have chosen the finest (or most interesting) 13 objects in this ANDROMEDA "GO TO" tour (there are 11 targets listed, but ONE GO TO for #11 - Messiers 31, 32 and ngc205); as with all GUIDES, all objects listed below will be visible in most telescopes (some naked eye) from a small refractor to a large reflecting telescope; of course larger apertures may "show" an object a bit closer and "better," but frequently a wide field and low power view is more desirable than aperture for FINDING the objects initially. Indeed, I strongly encourage you first FIND the target object, or its approximate location through your GO TO function with your lowest power and then - once IDENTIFIED positively - move up slowly in steps with magnification if necessary. Remember, not all objects "like" magnification. Sometimes better "field of view" (such as the wonderful wide fields provided by the modern APO refractor) is desired over light gathering (like large catadioptic scopes) and magnification.

The rule for determining "optimum magnification" is that: 1) too low power results in sky background glow detracting or diminishing the contrast against the deep sky object; 2) too high magnification darkens BOTH the sky background AND the object; 3) medium magnification can be achieved at which you have MAXIMUM contrast between the object and its darkened background sky. I have found through three decades of direct observing that about 15x per inch aperture (36x for the 3-inch); 55x for the 4-inch; 75x for the 5-inch; and, 125x for the 8-inch).for deep sky observing is PERFECT for most objects. That being said, always remember that DOUBLE or multiple stars require whatever power you can crank out....the seeing conditions are the limiting factor here.

For my complete and comprehensive discussion regarding seeing conditions and sky transparency, see my ASO GUIDE under the GUIDES/General tab on the Arkansas Sky Observatories website.

With all deep sky objects, avoid attempting to observe when the moon is in the sky, even a very thin crescent, as its brightness in the sky will overshadow the very dim contrast afforded by even the brightest deep sky object; if you see the object at all against moonlight, you will NOT see the subtle outlying areas or the full detail of what is presented.

Andromeda is a dominant constellation of fall skies. It is in that "quiet zone" that follows the spectacular summer Milky Way clouds and wealth of rich deep sky objects.....yet precedes the equally rich skies of winter, harboring the beautiful and bright constellations of Orion, Taurus, Cassiopeia and Canis Major. Andromeda rises almost due NORTHEAST for mid-northern latitudes about 10 p.m. local time during the first week of July. High in the north, it takes a while to reach the meridian, or highest point in the sky, culminating at 6 a.m. on that same night! However, midnight culmination - when it is on the meridian at midnight each year - occurs always on about October 5.

As with all of the "GO TO" tour constellation lists, I recommend a good star atlas and/or chart which will list all the finest objects, constellation-by-constellation. One very handy reference guide is the PETERSON FIELD GUIDE TO THE STARS AND PLANETS, which features complete lists with declinations, right ascensions, magnitudes, and all pertinent information for you to expand your observing horizons beyond this brief GUIDE.

The constellation tour Star Chart above will get you started on your journey for this constellation.

Following is the concise object list for your "GO TO" tour of ANDROMEDA; you may wish to find the majority of the objects from the AutoStar Library (for example, you can easily go to the WONDERFUL Messier 31 if you pull up "Object/Deep Sky/Messier/..then type in '31'...." and then press "Enter", followed by "GO TO" to access this naked-eye galaxy. On the other hand, if you want to experiment and become a "better AutoStar user" try entering the exact R.A. and DEC coordinates of that object as described above after holding down the MODE key. You will find the accuracy of entered GO TO's to be somewhat less than those stored in AutoStar, but the capability of acquiring unlisted objects is fantastic!

You will access your FIRST GOTO target – the brightest star in Andromeda, Alpheratz using the keypad menu or the tabs on your computer sky program. For Autostar: "SETUP / OBJECT / STAR / NAMED....and scroll to "**Alpheratz**"", then press "Enter" and subsequently "GO TO" to move your telescope to this bright star.

All sky programs have similar Named Star access; refer to the HELP tab on PC sky programs.

You may also similarly access the constellation by:

SETUP/OBJECT/CONSTELLATION/**Andromeda**
.....Enter....GO TO.

OBJECT 1:   bright star - ALPHERATZ (alpha Andromedae) - R.A. 00h' 06' / DEC +28 49 - Magnitude: 2.2

OBJECT 2:   bright star/double/galaxy - MIRACH (beta And) - R.A. 01h 07' / DEC + 35 21 - Mag. 2.0, great target!

OBJECT 3   double star - ALMACH (gamma And) - R.A. 02h 01' / DEC + 42 06 - Mags: 2.1 & 5.1 - VERY close!

OBJECT 4:   test double - 36 And. - R.A. 00h 52' / DEC + 23 21 - The "standard" for testing an 8" telescope!!

OBJECT 5:   great double for ETX 60 and all - 59 And. R.A. 02h 09' / DEC + 38 48 - Mags: 6 & 6.7 - wonderful star.

OBJECT 6:   Red Dwarf double! - Groombridge34 - R.A. 00h 16' / DEC + 43 44 - Mags: 8.1 & 10.9 - Really neat!

OBJECT 7:   variable/odd nova? - Z And.- R.A. 23h 31' / DEC + 48 32 - a star on a wild ride! Catch it while you can!

OBJECT 8: galactic cluster - ngc0752 - R.A. 01h 55' / DEC + 37 25 - Mag: 7.0, large, 70 stars over huge area!

OBJECT 9: - edge on spiral - ngc891 - R.A. 02h 19' / DEC + 42 07 - Mag: 11.8 - faint but "see-able" in A 3-inch+

OBJECT 10:   planetary nebula - ngc7662 - R.A. 23h 23' / DEC + 42 12 - Mag: 8.9, very nice green-blue

OBJECT 11: Andromeda galaxy, Messier 32, ngc205 - R.A. 00 40' / DEC + 40 00 - Three galaxies for one price!

A VISUAL GUIDE TO OUR DEEP SKY
OBJECTS IN ANDROMEDA

Object 1 - Our "Starting" Bright Star
"ALPHERATZ" (alpha Andromedae)
Also known as "Sirrah," this star originally was
designated by the early Arabic "namers of the stars"
as the "horse's naval," in reference to an early
association with the constellation of Pegasus, the
winged horse. It is quite easy to confuse this star as
one belonging to Pegasus since it marks the
northeast corner of "The Great Square" (see the
chart above) of Pegasus. Using this relatively bright
star has always been a benchmark for quick
allocating of the great Andromeda Galaxy for naked
eye observers. Alpheratz is at a distance of about
120 light years from Earth, making it a relatively
close neighbor.

Object 2 - *MIRACH* - (beta Andromedae) and NGC
404

Here is actually a "triple treat" in our Andromeda
guide. Beta Andromedae, actually brighter at
magnitude 2.0 than Alpha, is a double star for larger
amateur telescopes. "Mirach" is about half the
distance to us as Alpheratz and shows a true
companion of the 14th magnitude (should be visible
in the 8" scope under perfectly dark skies and
medium-high (about 200x) power nearly due south
(Position Angle 202 degrees) about 28" arc away
(that is fairly close, about half the diameter of
Jupiter would seem in the same eyepiece). It
"might" be glimpsed on very dark nights with a 6-
inch, as there are some estimates giving this faint
companion a magnitude of about 13.6. Easily

visible, but not bright, in the 5" scope will be NGC 404, located northwest of Beta only 6' arc! This will place it in the same field of view with the 26mm eyepiece and some of even higher magnification with that scope and an 8-inch. At 12th magnitude, this elliptical galaxy is too faint to be seen in smaller telescopes.

Object 3 - ALMACH - (gamma Andromedae) - A
Currently this double star is separated by nearly its maximum as viewed from Earth; with a period of 61 years, it passed periastron (most distant point from our perspective) in 1981, but still is located nearly 1" ENE of the brighter (magnitude 2.1) star. The companion is magnitude 5.1 and the color contrast between these two is fantastic....actually in smaller telescopes (you likely might see and elongation in a 4-inch, two "bumps" in a 6-inch and clear separation in the 8-inch) it is the contrasting COLOR that can separate the two stars....NOT a clear dark space! The brighter star is a distinct brilliant orange, while the secondary star appears green or green-blue, making a wonderful colored pair. Almach is 270 light years away.

Object 4 - The "Universal Standard Double Star Test" for a good 8 to 10" Telescope –
36 Andromeda
So you think you've got GOOD optics in that 6-8 inch telescope? Let's put them to the test. Some people say this star is "easy in a 6" reflector," while I have heard from others who cannot split the star with a classic 12.5 inch Cave Astrola Newtonian..

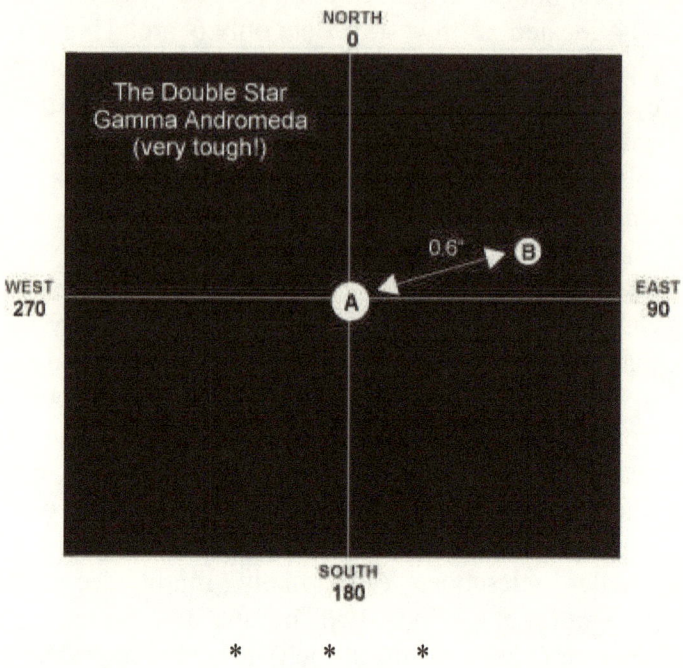

The Double Star
Gamma Andromeda
(very tough!)

*   *   *

Indeed, I have seen both stars in my 6" refractor under very steady skies, yet a 24" telescope fails to show them cleanly separated. This is a star that defies physics, predictability and logic. So I am not ABOUT to tell you if you can or cannot resolve the star. It should be resolved, but barely, in a 6-inch telescope and certainly if your optics are well collimated in an 8-inch. Presently the separation is given in many sources as 1.1" arc, but the star will seem much more difficult. The brighter star, 36 Andromedae at magnitude 6.2, is located at the same distance as Alpheratz from Earth; its companion - at magnitude 6.7 - is in true orbit around the star, and presently can be found nearly DUE NORTH of the slightly brighter primary; so look in a north-south orientation using very high magnification (about 50x per inch should be required). Overall the combined brightness of these two stars provides for a naked eye object of magnitude 5.4.

Object 5 - Here's One for Everybody! - 59 Andromedae - A Fine Double Star

This double star is sure to please everyone, although not as colorful and as bright as many others. The primary star is magnitude 6.0 and thus visible in small finderscopes; located nearly NE from that star is a slightly dimmer magnitude (6.7) star about 17" arc distant. Use medium to medium-high (about 15x to 25x per inch aperture) to see this pair the best. There IS a color contrast here if your eyes are good. The fainter star "might" appear very brilliant white or bluish white, while the brighter star should appear yellow or even yellow-orange...use higher magnifications to help bring out any color that you might "think" you see!

Object 6 - Here is a Wild One: "Groombridge 34 (GRB34)" - A Double Red Dwarf Star!
This one will most definitely be your "GO TO" USER OBJECT for Andromeda so you might go ahead and make a note of it. I chose it because it is remarkably interesting, both in fact and in vision. GRB 34 is the designation for a pair of RED DWARF stars - highly evolved "late" spectral stars of very small mass....as a matter of fact INCREDIBLY small mass. The larger star (magnitude 8.1) is only 1/6000 the mass of our own sun!! Think that is impressive? How about the smaller (magnitude 10.9) star, with its mass an astounding 1/40,000th smaller than the sun. Now

are you impressed? NOW do you see why I want this object on your "User Object" library list of the unusual and weird?

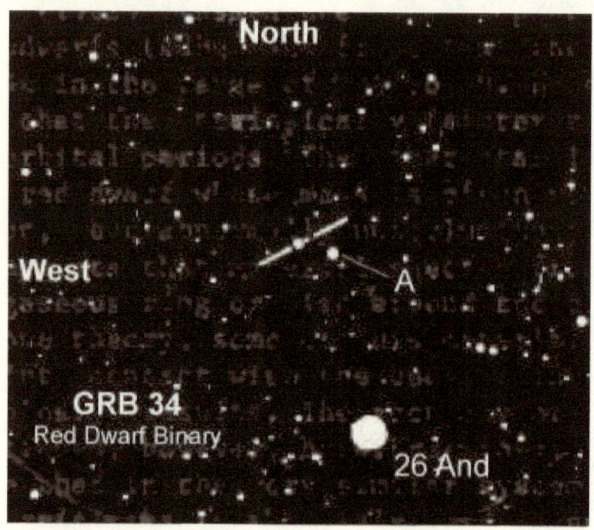

GRB 34 is actually visible in small telescopes, but you must know which star you are looking for in such a low power and wide field. Once a suspected star (look for a VERY red stellar object) is found in any scope, always begin slowly increasing the magnification until you finally are able to resolved into two components. This one should not be difficult, as the two stars are a full 39" arc apart, almost the same distance that Jupiter would fill in the same eyepiece. It is the faintness of the secondary that makes them difficult for discovery in the smaller telescopes.

Object 7 - In the sky!! Is it a bird....a plane....a variable star....a nova? Heck, we have no idea!
This star - Z Andromedae - is a most unusual variable and one that is frequently classified as a

46

"recurrent nova" by astronomers. You can make a career out of this unpredictable and erratic star.

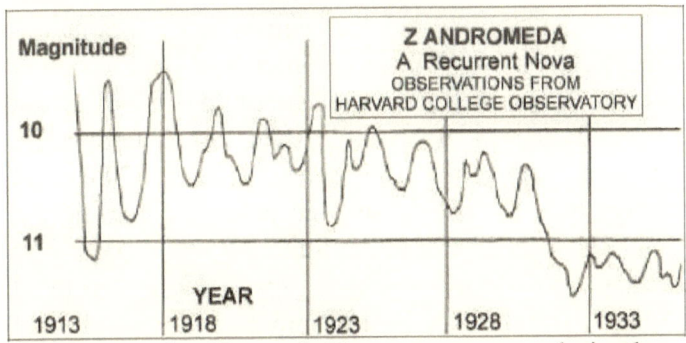

Looking at the Harvard Observatory records in the light curve above you can see that Z Andromedae reached a peak (BACK-AND-FORTH!) three times from 1913 through 191 and then began to slowly continue to peak....but a bit less each time. Major outbursts of the this star CAN be seen to have some type of regularity - about 695 days to be exact - but each time the star does NOT attain a brightness equal to the time before. Increases of up to two full magnitudes are dramatically rapid, followed by equally rapid declines! But notice what happened after about 1933 when the star decided it might be time to "rest" again.

To locate and monitor this wonderful variable star - suitable for long-period observing with the 3-inch and all telescopes larger - you merely log onto the web site of the *American Association of Variable Star Observers* (AAVSO) for valuable observing information and locator/comparison charts. As has been frequently mentioned in these "GO TO" TOURS, a variety of charts is always available for nearly every variable star known. For these charts, the AAVSO has selected very constant and stable comparison stars with exact magnitudes provided with which the variable can be compared. The "a" scale charts are naked-eye locator charts for brighter stars that required brighter comparison stars across the sky; "b" is slightly smaller scale with fainter stars; "c" is telescopic at medium magnification with stars typically visible in a 3-inch and larger scopes, while the "d" charts reach as faint as magnitude 14 and below. To assist in locating the star each time, refer to the link: https://www.aavso.org/apps/vsp/ . Note for these charts, simply type in the NAME of the variable at top to generate your choice of chart.

for the "a" chart (reversed for your catadioptic telescope field of view!). Then, once found, go the more narrow field and higher-power "g"chart to assist in detail comparison stars with specific confirmed magnitudes. I know that the AAVSO would greatly appreciate your weekly or monthly estimates of this unusual star's brightness changes!

Z Andromeda is actually a pair of stars, one red and one blue in the spectral range. Each is known by astronomers now as a variable in and of itself, both varying in synchronicity of some 700 days.....THAT in itself is a remarkable thing to ponder! Something to think about on the next cloudy night.

Object 8 - Galactic Cluster NGC 752 - A Very Nice LARGE Cluster, Ideal for a small APO!
Here's where the wide field refractors get all the other telescopes back....this beautiful but large cluster is all but lost in telescopic views; in fact, before quality small APO refractors were popular, either a large pair of good binoculars or a very expensive richest field refractor would have been

the instrument of choice to view this cluster. NGC 752 is much like other familiar clusters; at 1300 light years away, it is much farther and older than M44, the "Beehive" or Praesepe in Cancer, than the Hyades in Taurus (the closest star cluster to Earth at only 150 Light years) and the slightly more distant Pleiades, only 450 light years away. In spite of its brightness of 7.0, this is a very faint overall cluster because of its size, nearly 1.5 times larger than the diameter of the moon as we see it from Earth.

Nonetheless, of its 70-odd stars - all of which are visible in a telescope 6-8 inches, sixteen of them are easily viewed in the smaller telescopes, with magnitudes from 8.9 to slightly fainter than 10th. The nice wide field photograph taken with astrographic equipment at Lowell Observatory shows how effective the wide field of view can be. The field in this photo (above) is wider than is possible to attain with either the mentioned telescopes, yet most stars seen in this photograph ARE visible in both telescopes.

Object 9 - The Beautiful (but faint!) Edge-on Spiral Galaxy NGC891
This edge-on galaxy, though pretty, is NO rival to ngc4565 that I featured and discussed in detail in our "GO TO" TOUR of Coma Berenices. (see GUIDES/Constellation/Coma Berenices here at ASO). They both are very similar, yet this one in Andromeda, at magnitude 10.9 is much fainter and

more difficult to view; also it is smaller, at 11.8' x 0.8' arc, still very large for a deep sky object. However, remember that ngc4565 will literally stretch from one edge of a 32mm wide field Plossl to the other in a 6-8 inch telescope on a very dark night - almost one degree. This galaxy appears about one-half that extent, but is still a very nice object in the larger scopes. At least a 3-inch refractor is necessary for adequate views of this sliver of light. At a distance of 43 MILLION light years, this galaxy is believed to stretch 120,000 light years across, yet STILL is a smaller galaxy that the Great Andromeda Galaxy featured below! In the Mount Wilson 60-inch reflector photograph below, compare the similarities of this galaxy with ngc4565 at the link above. They are - photographically at least - remarkable similar; however visually there is a significant difference in the view. A very dark night and medium magnification is necessary to view this long shaft of light, regardless of which telescope is used.

Object 10 - A Lone Planetary Nebula, NGC 7662
This is a very difficult object, although its relative brightness given in all references (8.9) suggests otherwise. The overall size is "about" the diameter of Saturn's globe (not with rings) seen in the same eyepiece, so you can appreciate that it is relatively small. However, a 3-4 inch scope will definitely reveal its small disk-like appearance of blue-green ghostly glow. Larger telescopes may exhibit the annual or "ring-like" shape (see the drawing below by **E.E. Barnard** from 1907 through the great 40" Yerkes telescope - this drawing very closely matches visual impressions through an 8-inch, but with NO central star visible in the 8" scope).

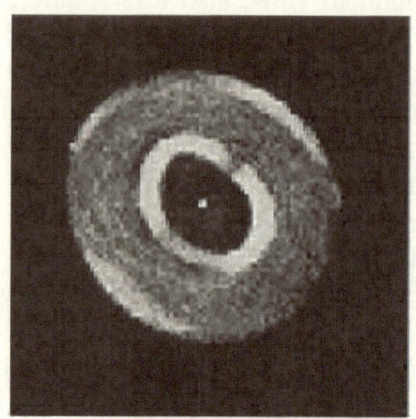

NGC 7662
A Bright Planetary
in Andromeda
DRAWING BY E.E. BARNARD
1907

Although the central star for this planetary nebula -
a shell of stellar gases emitted by the explosion of
an unstable star - is fairly bright (12.5 magnitude), it
is VERY difficult to see visually except in very
large telescopes; the reason is that there is much
extraneous gases that block the visible wavelengths
of light to the eye; averted vision in a 10-inch scope
will reveal the star but only with a magnification of
up to 400x and the darkest of nights. The star IS a
variable, changing in brightness unpredictably from
the 12th magnitude all the way down to 16th
magnitude....so if you cannot see it in an 8" or
larger scope: try again another night! In a 5" and
larger telescope, expect to see a true elliptical - not
circular - shape to this nebula; higher powers (about
30x per inch minimum) are recommended to see
both detail and the pretty blue-green glow of this
interesting planetary nebula

Objects 11 - Our Final Objects and Grande Finale!
The Galaxy Triplet: Messiers 31 and 32 & NGC 205!

There are memorable deep sky objects that are familiar naked eye targets. Which one is your favorite? Perhaps the best known is the Pleiades, or Messier 45 in Taurus, which many unfamiliar observers mistake for "the little dipper." Others include: Messier 44 (Praesepe) in Cancer, the Hyades in Taurus, the Great Orion Nebula in Orion and - of course - the Great Andromeda Galaxy, Messier 31 in Andromeda. As seen in the simple photography at the beginning of this guide, this object is a beautiful and striking reminder of just how vast the Universe really is and frequently serves to humble mankind through its vastness. Obviously, the fabulous near-twin to our own Milky Way, this galaxy has been known by mankind as long as it has been on this Earth. However, the earliest pseudo-scientific recording of the object was by the Persian star recorder *Al-Sufi* in about AD 905 who noted it as the "little cloud" in his Book of Fixed Stars; it must have been observed by Chinese and Persian astronomers even earlier. There have been many "multiple independent discoveries of the Great Andromeda Galaxy. Unaware of the Persian records of this object, the French comet hunter and "faint fuzzy" cataloger *Charles Messier* pretty much inferred that he was the first to actually pin down the coordinates of the object. who was the first to give a telescopic description in 1612. Unaware of both discoveries, *Giovanni Hodierna* independently "discovered" this object around 1654. *Edmund Halley*, however, gives credit for the discovery of this "nebula" to the French astronomer *Ismail Bouillaud* who observed it in 1661 even though *Bullialdus* mentions that it had been seen 150 years earlier (in the early 1500s) by someone else. I, too, have "discovered" this object so you can add my name to the list as well.

Simon Marius, in 1611, describes the "nebula" as "...the light of a candle shining...through horn," a very beautiful and poetic overview of just how it looks today.

The huge galaxy, in spite of all the "discoveries" is known as **Messier 31** today, or NGC 224. In a good 8-inch, you can clearly make out the dark dust lane in the northwest part, just away from the core, or "central hub" of the brightest part of the galaxy (see the photograph below). Best views of this by far are in very low power and wide field instruments, and thus the richest field refractors have a distinct advantage to this huge object.

The visual size of this object is phenomenal. Its length can be seen for a good 3 degrees (six lunar disks!) in a good pair of 10 x 50 binoculars on the darkest nights. It width (since it is tilted to our line of sight) is about equal to the diameter of the moon from Earth (about 35' arc). Although only listed as magnitude 4.1, the Andromeda Galaxy appears to be an object of much greater magnitude, simply because of its large extended size.

One of the largest spiral galaxies known to exist, and over 2.3 million light years away, Messier 31 is the closest spiral galaxy to our own Milky Way. The southern sky *Magellanic Clouds* are closer but are considered as irregular or peculiar galaxies, and unlike our own star system. As part of the "Local Group," this fabulous galaxy is teamed up gravitationally with others in the neighborhood including: Our Milky Way, both Large and Small Magellanic Clouds, Messier 33 in Triangulum, NGC 6822 (Sagittarius), IC 1613 (Cetus) and galaxies in what we describe as the "Fornax" and "Sculptor" system. In addition the TWO companion galaxies to the Great Andromeda Galaxy - Messier 32 and NGC 205 - are also members of the Local Group.

Messier 32 is a faint (9th magnitude) fuzzy appearing object that is clearly distinguishable in a 4-inch and larger scopes and is found very close (24' arc due south) from the central hub of Messier 31. It actually SMALLER than the other companion - NGC 205 - which is a large, diffuse blur in a small telescope and clearly shows. Both of these companions are elliptical galaxies.

Observers using larger telescopes should venture even further to find yet TWO MORE companion galaxies to the Great One. About 7 degrees north in Cassiopeia you will find NGC 185 (Mag. 10.3) and NGC 147 (Mag. 12.1) which are definitely observable in our larger scopes. You will find them listed in your computer program "Deep Sky" object library.

WANDERING ABOUT....YOUR NEW "USER OBJECT" IN ANDROMEDA

I alerted you to the fact that [Groombridge34 - R.A. 00h 16' / DEC + 43 44 - Mags: 8.1 & 10.9] would most definitely be our "GO TO" TOUR User Object for Andromeda. How can I resist...you already have white dwarfs, black holes, X-ray stars....why not a pair of red dwarfs to go with all this?

On AutoStar or your PC sky program, go to: "Select/Object [enter]...." scroll down to "User Object" [enter]. Now enter the coordinates given above for "GRB 34", using the number keys on AutoStar. After entering the coordinates and pressing "Enter" yet again, scroll down one and you can list the magnitude of the object as "8"[Enter].

Now you have yet another unique and oft-overlooked deep sky object for conversation-starters and crowd-stoppers at your next star party or club outing.

In Volume II of these GO TO" tours you will find another nearby close galaxy, that of **Triangulum**, and another in a series of "left-over stars" that some later astronomer - in this case the Frenchman *Lacaille* - decided should be grouped into a constellation of sorts simply because Ptolemy had not "used up" all the stars to his liking. Nonetheless, Triangulum - with its huge Pinwheel Galaxy, Messier 33, survived the final cut of the International Astronomical Union (IAU).

Surprising to most observers, this tiny constellation contains a wealth of great objects for modest and small telescopes! We will explore several nice galaxies and several beautiful double stars that await our "GO TO" constellation tour into Triangulum far into Volume II of *Constellations*.

Good Observing and may the stars serve as your sentries as you explore the frontiers of space!

*   *   *

Johannes Bayer's Andromeda, from his
1603 *Uranometria*

Chapter Two
AQUARIUS

This is the second Constellation Guide, "GO TO AQUARIUS" of the series for all GO TO telescope users. This constellation features MANY wonderful deep sky objects, interesting stars and much wonderful historical interest. Yet Aquarius is a notably "ignored" constellation by the amateur astronomer, even those in extreme southern latitudes.

I have always attributed this fact to the lack of bright "showcase" stars and at least one outstanding "trophy" object." Lyra has the "Ring Nebula", Andromeda has its galaxy, Taurus the Crab Nebula and the Pleiades, even lowly Vulpecula boasts the likes of the *"Dumbbell Nebula."*

If I ask you to name one outstanding visual deep sky object in Aquarius....can you do it? Likely not. Aquarius is located VERY low in the Summer northern skies, yet is perfectly placed for southern hemisphere and very low latitude northern hemisphere observers; as the sky chart shows, the **ECLIPTIC** - the band in which the sun, moon and planets all appear to move within - runs through this very large constellation, yet all objects within its boundaries will be both NORTH (positive "+" declinations) and SOUTH (negative "-" declinations) of the Celestial Equator which runs right through the northern regions of this large constellation.

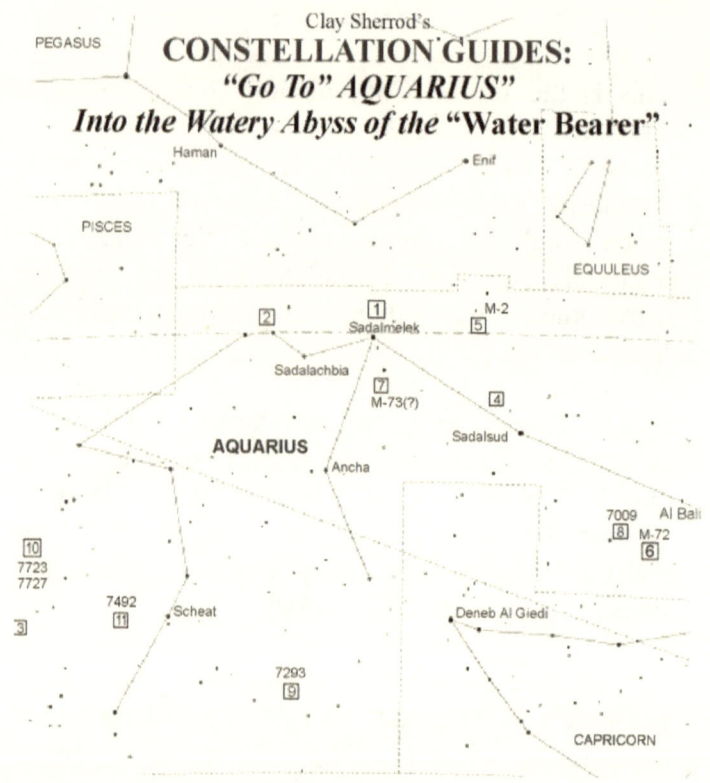

CONSTELLATION GUIDES:
"Go To" AQUARIUS"
Into the Watery Abyss of the "Water Bearer"

A finder chart for locating many of the GO TO objects in the constellation of Aquarius; if using a computer planetarium program, you are encouraged to plot the objects on your screen for higher resolution than this chart provides.

Aquarius is just ONE of seven constellations in this region that has some association with "water," and reveals the fascination of water subjects (perhaps due to arid conditions of the "original" Arabian namers of the stars?) to the earliest stargazers. In addition to the "aquatic" Water Bearer of Aquarius, we have: CETUS (a whale), PISCIS AUSTRINUS (southern fish), CAPRICORN (the "seat goat"), PISCES (a fish), DELPHINUS (the dolphin), and ERIDANUS (a celestial river). The importance of seafaring to the earliest people and the need for good navigational targets such as stars provide close to the southern horizons, likely gave rise to this "watery association."

In Greek legend it is from Aquarius, the "bearer of water" that the crow "Corvus" (west of Aquarius) was supposed to have stopped on its journey to apprise the gods of the losing battle against strong and perhaps immortal enemies to the west; but stopping to drink the water of Aquarius rather than telling the Gods of the ill-fated warriors' plight, the battle was lost by the refreshing detour chosen by the bird. Hence, for disobeying his orders and allowing the defeat of the Greek warriors, he is now eternally destined to fly at his given position of the sky with parched mouth, a dry rock in his beak and far from any refreshing relief from the Water Bearer.

Although the star field of Aquarius if far richer than that of Capricorn
(see Constellations/*Capricorn* in this volume) this area of our late summer skies is still fairly void of conspicuous star patterns that are memorable to the stargazer. It, for lack of better words, is a destination not often frequented by amateur astronomers. Nonetheless, there ARE some interesting objects in Aquarius that we will explore on this "GO TO" tour.

There are MANY galaxies in the realm of the Water

Bearer, but most of these NGC objects are too faint for our telescopes and, indeed, many are beyond the reach visually of even very the largest telescopes except photographically. Hence, these galaxies (and some other faint objects of other types) are not discussed in this guide. Throughout our "GO TO" guides, only objects that are within reach of telescopes ranging from 60mm to 200mm aperture are typically discussed; occasionally you will find noted an important object or "position" that is of historical or other interest that CANNOT be seen in our telescopes.

As with all "GO TO" tour guide star charts, use this accompanying chart as a map for locating the objects in this constellation visually and as a reference at the telesopce as you proceed through the Aquarius "GO TO" tour.

Also with every "GO TO" tour guide, each GO TO object in Aquarius is discussed for your telescope regarding the type of conditions necessary for you to view it optimally for discern the very faintest details.........magnifications and aperture necessary for most objects, and much, much more. This is YOUR complete guide to get you on your way to exploring the best (and few!) objects in these two constellations. The following listing of "BEST" objects contains the finest or most interesting from my own observing experience and preference.

Use the attached star chart and the following Guide as an excellent reference for your next star party itinerary, or a beginning for further study into the thousands of objects visible in this part of the sky. Truly these extensive Constellation study guides will most definitely put your telescope to work for you in the most efficient and enjoyable way possible! As a matter of fact, MANY GO TO users are now programming their own "Tours" based on these guides, using each constellation as a separate GO TO Tour for the AutoStar or PC sky program library that can be added in or deleted through the main edit screen on your PC or MAC computer.

We hope you enjoy these comprehensive guides to touring the constellations via your AutoStar or PC sky program and its computer-driven telescope. Each new installment is complete with diagrams, charts and illustrations that you will find nowhere else. Please let us hear YOUR feedback and your observations of each and every constellation after YOU have toured its vast reaches of our skies!

YOUR AQUARIUS CONCISE DIRECTORY OF INTERESTING OBJECTS –

Take time to familiarize yourself with the large constellation of Aquarius and its borders. Within this constellation, even though it may seem uninteresting from the "un-interesting wrappings" of its star-poor outline, there are many fascinating deep sky objects, the best of which are discussed here. Also there are MANY doubles stars and some of the nicest and most interesting variable stars of all. One of my favorite *"cataclysmic"* and unpredictable variables - AE AQUARII - is located here and affords a wonderfully bright erratic star that can change in the course of hours to observers using small and medium telescopes. That star is also detailed in this guide.

Aquarius is conveniently placed in late summer skies and lends itself well to very good and long-period observing for all telescope users both north and south of the equator. I rises in the east about dark (9 p.m. local time) on about August 1 and "*culminates*" (passes over the meridian at midnight) around August 25, remaining in the sky throughout that night, transiting the meridian just before midnight. All deep sky objects and difficult double stars are ALWAYS best observed when they are located nearly overhead (or as high in the sky as possible), thus requiring the observer to look through the thinnest portion of the Earth's "lens" of atmosphere and haze.

I have chosen the finest (or most interesting) 12 objects in this AQUARIUS "GO TO" TOUR; as with all guides, all objects listed below will be visible in most telescopes (some naked eye) from 3 to 8 inches; of course larger apertures may "show" an object a bit closer and "better," but frequently a wide field and low power view is more desirable than aperture for FINDING the objects initially. Indeed, I strongly encourage you first FIND the target object, or its approximate location through your GO TO function with your lowest power and then - once IDENTIFIED positively - move up slowly in steps with magnification if necessary. Remember, not all objects "like" magnification. Sometimes better "field of view" (such as the wonderful wide fields provided smaller telescopes) is desired over light gathering (like an 8-inch) and magnification.

The rule for determining "optimum magnification" is that: 1) too low power results in sky background glow detracting or diminishing the contrast against the deep sky object; 2) too high magnification darkens BOTH the sky background AND the object; 3) medium magnification can be achieved at which you have MAXIMUM contrast between the object and its darkened background sky. I have found through three decades of direct observing that about 15x per inch aperture for deep sky observing is PERFECT for most objects. That being said, always remember that DOUBLE or multiple stars require whatever power you can crank out....the seeing conditions are the limiting factor here.

For my complete and comprehensive discussion regarding seeing conditions and sky transparency, see my GUIDE on this subject under ASO GUIDES on the Arkansas Sky Observatories' website.

With all deep sky objects, avoid attempting to observe when the moon is in the sky, even a very thin crescent, as its brightness in the sky will overshadow the very dim contrast afforded by even the brightest deep sky object; if you see the object at all against moonlight, you will NOT see the subtle outlying areas or the full detail of what is presented.

With all of the "GO TO" tour constellation lists, I recommend a good star atlas and/or chart which will list all the finest objects, constellation-by-constellation. One very handy reference guide is the *PETERSON FIELD GUIDE TO THE STARS AND PLANETS*, which features complete lists with declinations, right ascensions, magnitudes, and all pertinent information for you to expand your observing horizons beyond this brief guide.

Following is the concise object list for your "GO TO" TOUR of AQUARIUS; you may wish to find the majority of the objects from the AutoStar Library (for example, you can easily go to the globular cluster Messier 2 if you pull up "Object/Deep Sky/Messier/..then type in '02'...." and then press "Enter", followed by "GO TO" to access this remote globular cluster. On the other hand, if you want to experiment and become a "better computer user" try entering the exact R.A. and DEC coordinates of that object as described above after holding down the MODE key. You will find the accuracy of entered GO TO's to be somewhat less than those stored in your sky program, but the capability of acquiring unlisted objects is fantastic!

OBJECT 1: brighter star - SADALMELEK (alpha Aquarii) - R.A. 22h 03' / DEC (-) 00 34 - Magnitude: 3.2

OBJECT 2: good double star - Zeta Aquarii - R.A. 22h 26' / DEC (-) 00 17 - Mag. 3.7, good star for scopes

OBJECT 3: nice variable star - R Aquarii - R.A. 23h 41' / DEC (-) 15 34 - Mag. range = 5.8 to 11.5 - erratic, nice!

OBJECT 4: wonderful variable - AE Aquarii - R.A. 20h 38' / DEC (-) 01 03 - A very rapid irregular variable, nice object

OBJECT 5: very concentrated globular - Messier 2 (ngc7089) - R.A. 21h 31' / DEC (-) 01 04 - Mag: 6.3 - Super!

OBJECT 6: faint globular - Messier 72 (ngc6981) - R.A. 20h 51' / DEC (-) 12 44 - Mag: 9.8, small & unresolveable

OBJECT 7: a Messier MISTAKE! - Messier 73

(ngc6994) - R.A. 20h 56' / DEC (-) 12 51 - Did Messier goof up??

OBJECT 8: "Saturn Nebula" - ngc7009 - R.A. 21h 01' / DEC (-) 11 34 - Mag: 8.4, very interesting in larger scopes

OBJECT 9: "Helix Nebula" - ngc7293 - R.A. 22h 27' / DEC (-) 21 06 - Mag: 6.5 - very larger - ideal scope is low power!

OBJECT(s) 10: two close galaxies - ngc7727 & ngc7723 - R.A. 23h 36' / DEC (-) 13 14 - Mags: 11.1 & 10.7, nice

OBJECT 11: faint globular - ngc7492 - R.A. 23h 06' / DEC (-) 15 54 - Mag: 10.8, very faint, star-like in small scopes

A VISUAL GUIDE TO OUR DEEP SKY OBJECTS IN AQUARIUS

Object 1 - Our "Starting" Brighter Star - "SADAL MELEK" (alpha Aquarii)
This giant star may be up to 80 times larger than our own sun, but at a distance of 1100 light years, it appears only as a 3rd (2.92) magnitude star. It has the same spectral identity as the sun ("G"), but is huge by comparison, and over 6,000 times brighter if placed at the same distance! The wide-field view surrounding this star is a pretty sight on a very dark night, particularly with the wide field of view afforded by quality APO refractors.

Object 2 - Zeta Aquarii - A Wonderful Double Star for the smaller telescope
This is a wonderful double star for a 3-inch as a test object, only about 1.6" arc separating its two components. It should be a relatively easy object in a 6 to 8-inch telescope, with the brighter component

being magnitude 4.4 and the secondary only slightly more dim at magnitude 4.6. The two orbit one-another every 600 years or so and the closest the seem to get (right about now!) is the current separation that we witness. Since the theoretical resolution limit of the 3-inch is 1.3" arc, it should easily resolve this double with pretty high magnification (around 200x) and steady air. Currently look for the two stars in a nearly north-south orientation to one-another, a perfect alignment for a perfect test in small telescopes!

Zeta is the central star in the familiar asterism known as the "Water Jar" of Aquarius, a Y-shaped pattern formed by Gamma, Zeta, Pi and Eta in this constellation as shown in the chart below.

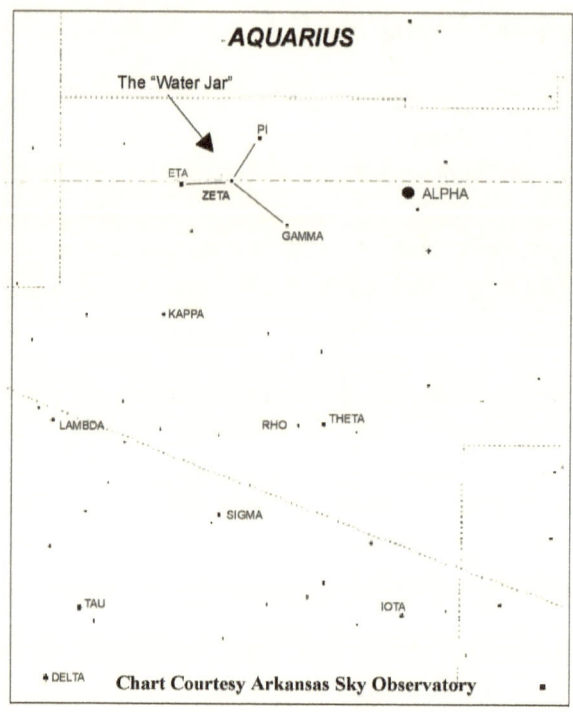

Object 3 - A Great Variable Star - **R Aquarii**
There is no wasted time when observing this wonderful and erratic variable star! There is so much to this complex star that an entire evening could be spent observing and contemplating the nature of this curious object. R Aquarii is a huge Red Giant pulsating variable star which likely went "nova" about 600 to 700 years ago and expelled a huge cloud of gas and dust, a shell that can be photographed, but not seen as attested in the wonderful photo below from the 100" Lick Observatory reflector.

THE "BAT NEBULA"
surrounding the variable star
R AQUARII

Coined the "bat nebula" from its characteristic shape, this is a rapidly-expanding cloud of gas that was ejected during the star's explosion. At present, the star is classified as a "semi-regular" star because it exhibits a rough "period" of about 386 days during which it might brighten to magnitude 6.0 or thereabouts. The light curve shown following demonstrates that this star is ANYTHING but predictable, and that it is known for long periods of time in which rapid and minor fluctuations around magnitudes 8 and 9 occur.

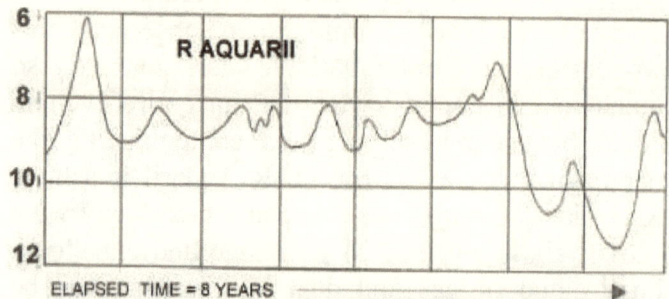

ELAPSED TIME = 8 YEARS ─────────────────►

To locate and monitor this wonderful variable star - suitable for long-period observing with and all telescopes - you merely log onto the web site of the American Association of Variable Star Observers (AAVSO) for valuable observing information and locator/comparison charts. As has been frequently mentioned in these "GO TO" TOURS, a variety of charts is always available for nearly every variable star known. For these charts, the AAVSO has selected very constant and stable comparison stars with exact magnitudes provided with which the variable can be compared. The "a" scale charts are naked-eye locator charts for brighter stars that required brighter comparison stars across the sky; "b" is slightly smaller scale with fainter stars; "c" is telescopic at medium magnification with stars typically visible in the 3-inch and larger scopes, while the "d" charts reach as faint as magnitude 14 and below.

For R Aquarii, access the chart https://www.aavso.org/apps/vsp/ . Note for these charts, simply type in the NAME of the variable at top to generate your choice of chart, save it to file, open and resize to print for the "a" finder chart that will get you started. Since R Aquarii rarely gets too faint, for comparison purposes the chart will serve for all telescopes as an ideal source for good comparison stars for the curious cycle of this star.

Object 4 - AE Aquarii - A Fantastic Variable for our telescopes

Using the same source, you can access the star charts for AE Aquarii, another wonderful and very unpredictable "cataclysmic variable" star in Aquarius. Locate the star first using the "a" chart https://www.aavso.org/apps/vsp/ . Note for these charts, simply type in the NAME of the variable at top to generate your choice of chart - and then move up to the very detailed and fainter limiting magnitude chart "g" to make your comparisons as AE Aquarii takes you through its roller-coaster fluctuations on a minute-by-minute basis. In 1982 I was involved in a major study of cataclysmic variable stars such as this one; it required monitoring 32 stars, each at a 30-minute interval over an entire night! In the case of AE Aquarii, visual changes in this star amounting to 0.3 magnitude could be seen in a period of only hours, creating a feeling of true change in the Universe!

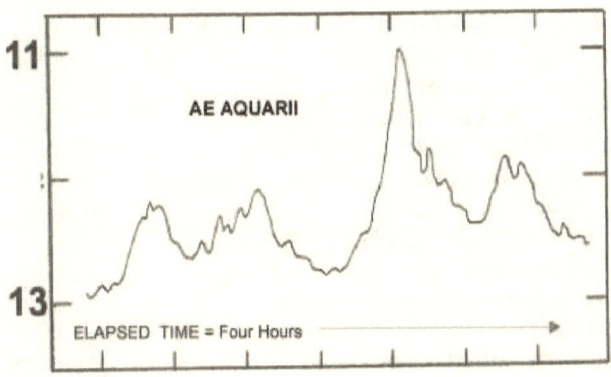

There is CONSTANT change going on with this

star....it will "sit" idle for a period of time, followed by sudden and explosive changes in brightness that can actually double the magnitude of this star! Very rapid changes that can be detected through our telescopes can be seen in less than one hour! Normally the flare-ups that characterize this interesting star will be followed by a somewhat slower (but still fairly fast) decline in brightness, only to have another follow in just a few days....like a "recurring 'mini-nova' ." At a typical 9th or 10th magnitude this star is easily followed in 3-inch and larger scopes. In actuality, it is a binary star of which the two components are LIKELY separated in space with LESS DISTANCE that there is between the Earth and its Moon! That is fantastically close for two stars. Thus, the outbursts are thought to be result of very volatile gases being "pulled" from a cooler star to a very HOT star adjacent to it, at which time the "new" gases add fuel to the fires and result in flare-ups.....much like squirting lighter fluid on your already-burning charcoal grill.

Object 5 - A Great Globular Cluster - Messier 2 - (It's been a while since we've seen a good one here!) Finally! A Messier Deep Sky Object that we don't have to squint to see! Messier 2 is oddly located nearly one-half the way around the sky from Messier 1. Go figure! At any length, this is a very nice globular cluster, although not on the same scale as the famous Messier 13 in Hercules, or those in Ophiuchus. For a complete observing guide to globular clusters, see my description at:

Messier 2, at magnitude 6.3 will appear distinctly as something "other than a star" in very small scopes and gets better with increased aperture; even at the very lowest power and widest fields, this is a nice object, appearing as a faint round glow in all instruments; increased magnification in the larger instruments will begin to show many delicate stars along the outer edges. This cluster, unlike Messier 3 and Messier 13 is very compact and concentrated; it is only about 7' arc across visually, compared to 11' for Messier 13. This is one of those deep sky objects where a bit of increased magnification will assist in a better view; I recommend using about 20x per inch for ideal viewing. Although over 100,000 stars populate this globular its vast distance of over 50,000 light years - much farther than M-13 - makes for difficult resolution. This huge object would be spectacular if placed at the same distance as Messier 13.

Object 6 - Another Globular Cluster.....Messier 72 - Not quite so pretty as Messier 2!
You might want to keep Messier 2 handy for comparison with Messier 72, another globular cluster in Aquarius. Just 3 degrees NE of M-72 is another object on our TOUR, the "Saturn Nebula", ngc 7009. Although the visual magnitude of M-72 is 9.1, bright enough to be a relatively easy object in a 4-inch scope, its size makes for a difficult object, at only 4' in diameter. This will appear as a very fuzzy star in the smaller instruments. The 8-inch, with powers up to around 300x, can begin to see some indications of stars around the outer edges of this small, compact and very round cluster. Nonetheless, it is a good object for all telescopes, noting that you can brag that this is "....one of the most difficult globular clusters of any to resolve" in amateur telescopes, and is an incredible 60,000 light years away, far more distant that even Messier 2,

discussed above. Incredibly, there have been over 45 variable stars actually discovered and monitored within this very distant object!

Object 7 - The Messier Object that "Isn't" - Messier 73
Either Charles Messier actually DID see a comet in October of 1780 when he logged this position of his 73rd "faint fuzzy" (he was, after all, making a list of all objects that MIGHT be confused for faint, non-tailed comets!), or he was suffering a setback after attempting to resolve stars in his 72nd object, discussed above. For at this position - which you really need to take a look at - there is NOTHING! Four stars, magnitudes 10.5 (two of those), 11.0 and 12.0 all together. Keep in mind that his small telescope likely might not have been able to detect the fainter star, so the mysterious nature of WHY he cataloged an object here is even more compounded. Messier recorded "3 or 4 stars" here, but in addition noted that there was nebulosity associated with those stars.....had he, indeed, seen a COMET superimposed OVER those distant stars? It is entirely possible and if so, he likely was the only living person to have ever seen that one, for it was never verified and there is no way to determine if it was seen before or after his sighting of whatever occupied the space now known as "Messier 73!"

Object 8 - THE SATURN NEBULA - NGC 7009
This is a very unusual planetary nebula, not at all appearing nor physically matching the "typical" planetary nebula like the Ring Nebula (M-57, see ASO *Constellations/Lyra* in Volume II) or the large "Helix Nebula" discussed for Aquarius (below).

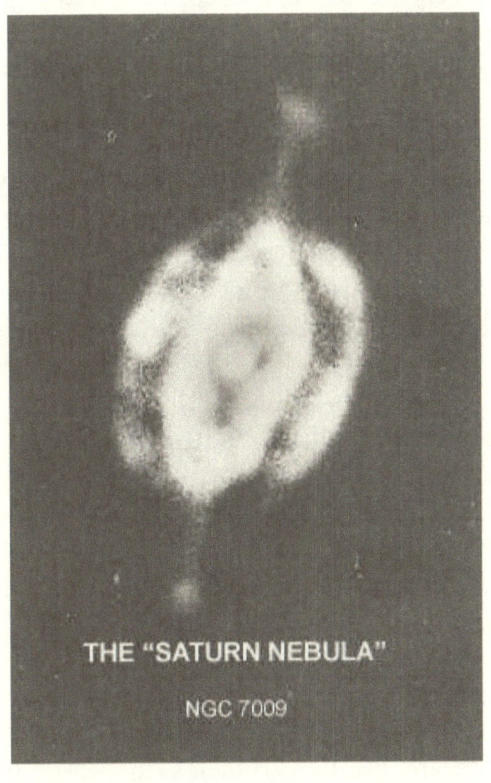

THE "SATURN NEBULA"

NGC 7009

The "Saturn Nebula" is so-named because of its two gaseous arm-like extensions (or "ansae") which project nearly north-south from the center of the gas cloud. This nebula is small (44" x 26", about the size of Jupiter's disk) and bright (8.4 magnitude, easy enough to see in a small APO refractor with about 60x +), so it is a good object for all telescopes. Medium, to medium-high magnifications are recommended for best views. In the 6-inch and larger telescope this is an OUTSTANDING object, rivaled in mystique only by "the ghost of Jupiter" (for computer see "Object / Deep Sky / Named....."). There is a distinct greenish or blue-green color to ngc7009 that almost looks falsely-lit in the telescope! In the eight inch the ansae can be clearly seen with about 225x, and on very, very dark nights in the six inch as well; in smaller telescopes, only the brighter central planetary nebula shell is visible. This is actually a fairly close planetary nebula to Earth, at only 3900 light years distant. The magnitude 11.7 central star which created the Saturn Nebula may be glimpsed on very dark nights at about 150x with a 4-inch and certainly with telescopes 6 inches and larger.

Object 9 - Another Wonderful Planetary Nebula - The Very Large "Helix Nebula"
The largest such object of its kind, the "Helix Nebula, ngc7293, is nearly half the size of the lunar disk - about 16' arc across! It has a total brightness (if all of its spacious size was compressed into the size of a star's disk) of 6.5, making it a very bright object. However, because that brightness is spread over such a large area, it is best seen in very low power, wide field views. Hence, the small APO refractors are ideal instruments for observing this beautiful "ring" of stellar gases. In larger telescopes, be sure to use your lowest magnification and widest field of view possible. This object can clearly be seen as a small round glow in good 7 x 50 or 10 x 50 binoculars. Although not appearing as bright, the Helix Nebula appears very much visually (and photographically) as the famous Ring Nebula in Lyra, as attested in the 200-inch Palomar photograph below:

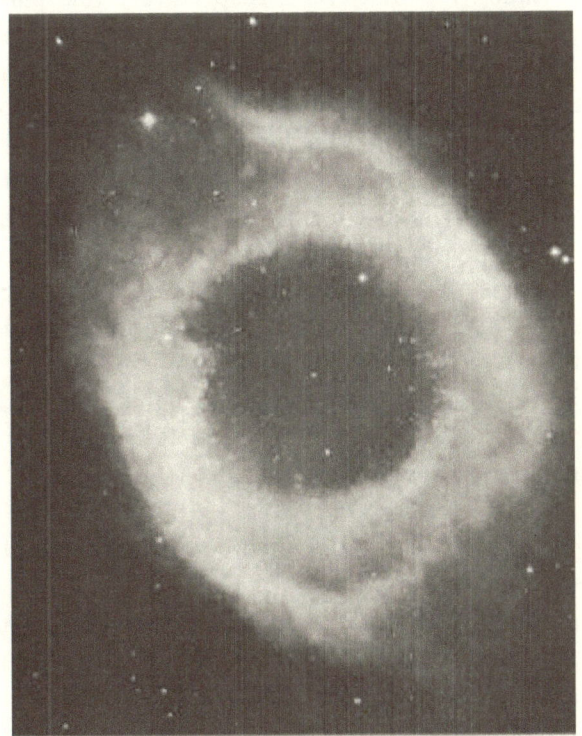

THE BEAUTIFUL HELIX NEBULA
NGC 7293 in Aquarius

In addition to being the brightest and largest of its type, ngc7293 is also maybe the CLOSEST to Earth, at only 85 light years! As a matter of fact the close star which "nova-ed" to produce the shell is still visible in perhaps even in a 6-inch telescope as a very central 12.5 magnitude star. This star is clearly visible in an 8-inch or larger telescope.

Objects 10 - Two "Companion Galaxies": ngc 7723 and ngc7727
These are two galaxies that are within the same low power, wide field of view, even in a good 3-inch. At

magnitudes 11.1 and 10.7, respectively this pair is often overlooked when some attention is really deserved of the two. Use very low power to find and view them together, separated only by about 45' arc. Both are spiral galaxies with no discernable features in any of the telescopes, but beautiful nonetheless within a fairly rich field of faint stars. ngc 7723 measures a small 2.2' x 1.6' arc oval while ngc7727 is nearly round and about 3' arc across. With the LX 90, you MIGHT be able to sometimes glimpse two delicate and faint "arms" extending up to 5' arc from the central region of this galaxy. Do not expect to see either of these faint galaxies in small, low powered telescopes.

Object 11 - Our Final Object - the faint globular cluster ngc7492
Sometimes given as magnitude 10.8 and others a much fainter "12.3", this globular is easy enough that I tend to agree with the former estimate. Either way, it certainly is NO M-2. It is larger however, than Messier 72 in Aquarius, at a diameter of nearly 3.5' arc, still nearly three times smaller than Messier 2. At this small size, it is NOT visible in telescopes under 4-inches, but can be detected looking like a faint nebulous star in a 6-inch and perhaps a bit better in the 8-inch. This globular is nearly twice the distance from Earth as Messier 2, hence its small size and faint apparent magnitude.

WANDERING ABOUT....YOUR NEW "USER OBJECT" IN AQUARIUS

How can I resist....was there EVER any doubt that astronomers make mistakes? (....or is there any doubt that we don't ALWAYS make mistakes?). If you believe in cosmic perfection, then believe that there is, indeed, nebulosity at position R.A. 20h 56.2m, DEC -12 degrees 51m, the position of Messier's 73rd object. You can impress your friends and neighbors with the fact that they are seeing something that Messier did NOT see: nothing. It is really a very interesting conversation point....Did Messier, over two centuries ago, actually discover a COMET at this position and NOT KNOW IT? It is entirely possible, and you can show everyone where it actually happened (or should that be "didn't happen?"). This missing Messier object should definitely be part of your USER OBJECT library!

On your GO TO system, go to: "Select/Object [enter]...." scroll down to "User Object" [enter]. Now enter the coordinates given above for "Messier 73", using the number keys on AutoStar. After entering the coordinates and pressing "Enter" yet again, scroll down one and you can list the magnitude of the object as "0"[Enter].

So with the addition of the "missing Messier object" (there are actually "two" - the other has been found, thank you) - you are completing a set which includes a black hole, dark nebulae, curious multiple stars, variable stars and weird deep sky objects. All of these are wonderful for conversation-starters and crowd-stoppers at the big astronomical events!

*   *   *

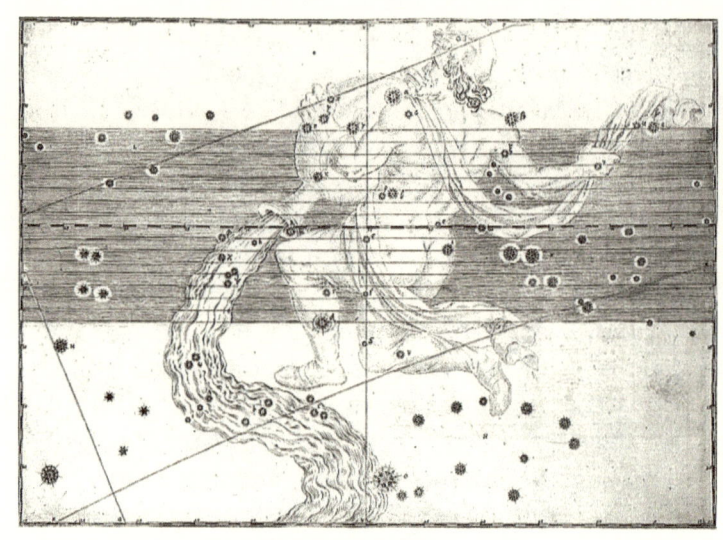

Aquarius, the Water Bearer
From J. Bayer's *Uranometria*, 1603

Chapter Three
AQUILA - SCUTUM

*The Milky Way's Soaring Eagle of the Night / The
Warrior's Lost Shield*

In our next Constellation Guide, "GO TO
AQUILA" of the series for computer telescope
users we will soar with eagles throughout the "great
river" of the Milky Way galaxy. Aquila and its
bright star ALTAIR are the southernmost of the
dominant summer constellations and together with
Deneb (Cygnus - see *Constellations* and Vega (Lyra
-also in Constellations, Vol. II) form the bright
"*summer triangle*" so unmistakable in our northern
skies. In addition to Aquila in this part of the sky,
we find more southerly SCUTUM (the "shield"), a
small but remarkably rich constellation often
overlooked or mistakenly thought to be part of
larger Aquila. The stray shield is that beaconing the
warrior Hercules, who has lost his protective barrier
in the many challenges he faced throughout Earth
and sky.

It is through Cygnus and down into Aquila that the
brilliant Sagittarius arm of our Milky Way galaxy is
seen; when we gaze at this wonder, we are looking
across a vast emptiness of space toward a star-and-
nebula rich spiral arm of the galaxy, and the deeper
we go toward the rich star clouds of Sagittarius, the
closer we peer at the very hub of the incredible
Milky Way and its **200 billion** some-odd stars.
Image that the earth is a planet on but one isolated
star in an OUTER ARM of the galaxy.....as we gaze
toward Aquila and Sagittarius, we are looking

INWARD toward the nuclear hub of the galaxy of which WE are part. Looking the opposite way - toward Orion and Auriga in winter skies, we look in yet another direction and at yet another, less star-dense galactic arm leading OUT OF the spiral system of stars.

The next closest galaxy that "resembles" our own Milky Way galaxy is nearly 2.5 billion light years distant. Every star, planet, cluster, comet, asteroid, meteor, globular...that you see is in OUR galaxy; once beyond all that "stuff" there is NOTHING....barely a molecule, until you reach the confines of the Andromeda Galaxy. Imagine yet further that - as you enter that galaxy some 2.5 billion years from now as you travel at "*warp one*" (the speed of light) - you begin to see new stars, nebulae and all those things similar to that same stuff from our galaxy that you left behind. Yes, indeed, the "stuff" of which we are made of is all the same.....universally everywhere.

Although bordering the star-rich Milky Way and containing many wonderful star fields for low power and slow scanning on a dark night, Aquila is curiously void of spectacular deep sky objects and remarkable multiple stars that are noted in such great numbers nearby. Even the most famous object nearby is often mistakenly placed within the confines of the large eagle's outstretched wings and talons, but indeed the "*Great Scutum Star Cloud*" with its fantastically rich star cluster Messier 11 is nonetheless still in "Scutum" to Aquila's south.

Hence, we will include some discussion of that minor constellation here since one of the most remarkable regions of stellar density of our Milky Way is seen within its boundaries.

Clay Sherrod's
CONSTELLATION GUIDES:
"Go To" AQUILA - The "Soaring Eagle".
also with discussions on SCUTUM

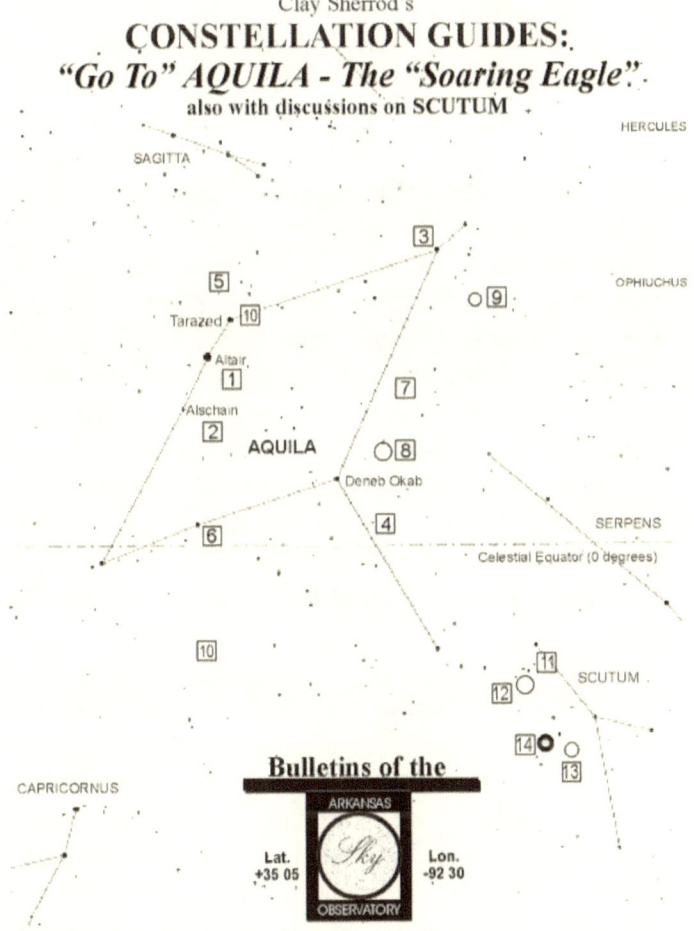

A finder chart for locating many of the GO TO objects in the constellation of Aquila and Scutum; if using a computer planetarium program, you are encouraged to plot the objects on your screen for higher resolution than this chart provides.

Note from the sky chart included here that the *CELESTIAL EQUATOR* passes through the middle sections of Aquila and just north of tiny Scutum. This is the reading "0" degrees on your properly adjusted declination setting circle. All angles NORTH of this equatorial line are positive ("+") and all angular measures (declinations) south of the celestial equator are negative ("-"); hence you will see references in this "GO TO" guide to both "+" and "-" declinations for celestial objects.

You will begin your "GO TO" journey into the Eagle's lair and into the great lost shield of the sky warriors via the bright star "**ALTAIR**", a nice bright yellow star that is commonly referred to in the "asterism" known as the SUMMER TRIANGLE, a nice wide shape bounded by the bright summer stars Deneb (Cygnus), Vega (Lyra) and Altair (Aquila). See our constellation "GO TO" tour for Cygnus at our Guides/Constellations here in Volume I) for a sky chart showing this spectacular summer marker!

Each GO TO object is discussed for your telescope regarding the type of conditions necessary for you to view it optimally for discern the very faintest details....double star challenges for each size telescope .....magnifications and aperture necessary for most objects, and much, much more. This is YOUR complete guide to get you on your way to exploring this large and interesting constellation. ONE THING that I might add about the region in and surrounding Aquila: Although some interesting and "test" double stars are always included in the "GO TO" tours, there is a particularly wide range of very interesting double and multiple stars too numerous to list or mention here. Consult a good handbook, such as the *"Burnham's Celestial Handbook,"* Vol. 1 for a very comprehensive list of locations, magnitudes and angular separations of these wonderful stars. There are many stars for EVERY telescope size and type.

Use this attached star chart and the following Guide as an excellent reference for your next star party itinerary, or a beginning for further study into the thousands of objects visible in this part of the sky. Truly these extensive Constellation Study Guides will most definitely put your AutoStar or other telescope program to work for you in the most efficient and enjoyable way possible! As a matter of fact, MANY telescope users with GO TO capability are not using smart phones and pads to actually link wirelessly via their devices to operate the entire telescope via a touch screen!

We hope you enjoy these comprehensive GUIDES to touring the constellations via your AutoStar or sky program and its computer-driven telescope. Each new installment is complete with diagrams, charts and illustrations that you will find nowhere else. Please let us hear YOUR feedback and your observations of each and every constellation after YOU have toured its vast reaches of our skies!

YOUR AQUILA / Scutum CONCISE GO TO Guide

There are literally thousands of fantastic objects in these two constellations, ranging from multiple stars, clusters, bright AND dark nebulae - from naked eye, to wide field, to telescopic - that is was very difficult to select (and limit) the number of objects for our "GO TO" TOUR. Only the best and in some cases, most challenging, objects are chosen for this brief tour.

I have chosen the finest 14 objects in this Aquila and Scutum to "GO TO" tour; as with all guides, all objects listed below will be visible in all telescopes (some naked eye) from the 3-inch through 8-inch telescopes; of course larger apertures may "show" an object a bit closer and "better," but frequently a wide field and low power view is more desirable than aperture. This is the case for MANY of these objects since we are looking directly into the very star-cloud-rich areas of our Milky Way galaxy. Indeed, I strongly encourage your to step away from the telescope often and scan the beautiful open skies and star fields with a good pair of 7 x 50 or 10 x 50 binoculars. You will be tempted to venture to your "dark sky site" for a full evening of laying back on a blanket and scanning the skies. The many deep sky objects in the Cygnus area even stand out to the dark-adapted naked eye, and the dark and bright star clouds emerge like in no other area. Once your eyes are fully dark adapted, you will even be able to see such wonders perhaps as the magnificent and very large "*Scutum Star Cloud*"....appearing as its namesake a very open and bright cloudy object that is huge in the open sky, sprinkled with thousands of tiny naked eye stars. It is within this magnificent cloud that the famous "*Wild Duck Cluster*" (really don't like that name for such a "royal" sight!) is embedded. Even the binoculars will show dozens of deep sky objects that cannot be fully appreciated in large telescopes with limiting fields of view!

The convenient sky placement of the large Aquila constellation lends itself well to very good and long-period observing for observers both north and south of the equator. When rising about dark in the east (mid-spring) it will remain in the sky throughout the night, transiting the meridian at about midnight during that season. All deep sky objects and difficult double stars are ALWAYS best observed when they are located nearly overhead (or as high in the sky as possible), thus requiring the observer to look through the thinnest portion of the Earth's "lens" of atmosphere and haze.

As with all of the "GO TO" tour constellation lists, I recommend a good star atlas and/or chart which will list all the finest objects, constellation-by-constellation. One very handy reference guide is the *PETERSON FIELD GUIDE TO THE STARS AND PLANETS*, which features complete lists with declinations, right ascensions, magnitudes, and all pertinent information for you to expand your observing horizons beyond this brief guide.

Following is the concise object list for your "GO TO" tour of Aquila and Scutum; you may wish to find the majority of the objects from the AutoStar Library (for example, you can easily go to the "Wild Duck Cluster" if you pull up "Object/Deep Sky/Messier Object/..type in '11'...." and then press "Enter", followed by "GO TO" to access my favorite beautiful galactic cluster. On the other hand, if you want to experiment and become a "better computer user" try entering the exact R.A. and DEC coordinates of that object as described above after holding down the MODE key. You will find the accuracy of entered GO TO's to be somewhat less than those stored in the computer, but the capability of acquiring unlisted objects is fantastic!

OBJECT 1: very bright star - ALTAIR (alpha Aquilae) - R.A. 19h 48' / DEC + 08 44 - Magnitude: 0.8, very near star

OBJECT 2: very tough test double - ALSCHAIN (beta Aquilae) - R.A. 19h 53' / DEC + 06 17 - Mags: 3.9 & 11.4

OBJECT 3: LX 90 test star - Zeta Aquilae - R.A. 19 03' / DEC + 13 47 - Magnitudes: 3 & 12 (!) very tough double

OBJECT 4: 3-inch test star - 23 Aquilae - R.A. 19h 16' / DEC + 01 00 - Magnitudes 5.5 & 9.5 - Test for 4-inch!

OBJECT 5: Nice Double! - Pi Aquilae - R.A. 19h 46' / DEC + 11 41 - Magnitudes: 6 & 7 - good double for 4-5 inch scope

OBJECT 6: Variable star - Eta Aquilae - R.A.19h 50' / DEC + 00 52 - Naked eye/ETX 60/70, Mag. range 3.5 to 4.5

OBJECT 7: Wonderful variable! - R Aquilae - R.A. 19h 04' / DEC + 08 09 - Magnitude range - 6 to 11.7, 300 days!

OBJECT 8: Small galactic cluster - ngc6755 - R.A. 19h 05' / DEC + 04 09 - Mag: 8.3, about 50 stars!

OBJECT 9: another galactic cluster - ngc6709 - R.A. 18h 49' / DEC + 10 17 - Mag: 8.1, nice, 30 stars

OBJECT 10: wonderful DARK nebulae - B143 & B133 - R.A. 19h 38' / DEC + 11 00 - (for B142 in the "great rift")

OBJECT 11: (in Scutum) Great variable star - R Scuti - R.A. 18h 45' / DEC (-) 05 46 - Mag. range 5.8 to 7.9, GREAT for small scope

OBJECT 12: (in Scutum) "Wild Duck Cluster" - Messier 11(ngc6705) - R.A. 18h 48' / DEC (-) 06 20 - Gorgeous 200 stars!!

OBJECT 13: (in Scutum) galactic cluster- Messier 26 (ngc6694) - R.A. 18h 43' / DEC (-) 09 27 - Mag.
9.3, 20 stars, nice/small.

OBJECT 14: (in Scutum) globular cluster - ngc6712 - R.A. 18h 50' / DEC (-) 08 47 - Mag. 8.9, very small & interesting

A VISUAL GUIDE TO OUR DEEP SKY OBJECTS IN AQUILA and Scutum

Object 1 - Very Bright Star - "Altair" (alpha Aquilae)

As usual, we are starting out Aquila/Scutum "GO TO" tour with the brightest member of the group. Altair, although uninteresting except for its brilliance in white-yellow grandeur, is a very peculiar and close star to our own sun. At magnitude 0.8, it is the 12th brightest star in the sky, partly because it is so close at only 16 light years. In many respects Altair is like our sun, but in MAJOR areas it differs greatly!

Altair (the "Eagle's Heart") is one of the fastest rotating objects known. It spins on its axis (making a rotation every 6.5 hours compared to our sun's 25 DAYS!) faster than even speedy Jupiter (9h 55m), and so fast that its sphere is actually flattened greatly from the spin, like an oblate or flattened basketball. It is about the same size as the sun and has an OPTICAL companion star of magnitude 10.1 (not a physical double) that might be seen with difficulty (due to the brightness of Altair itself) nearly due north (and a bit west) of the bright star. It is quite a distance, about four times the disk diameter of Jupiter as seen in the same eyepiece. Do not bother with any scope smaller than a 6-inch as the faint companion will not be seen. Luck and imagination MAY give you a glimpse in the larger scopes.

Object 2 - "Alschain" - (beta Aquilae) - A Very Tough Double for the 6-8 inch telescope
Although Alschain is a widely separated double star (12" arc) it is still a tough object because of the brightness of Beta Aquilae (magnitude 3.7) and the VERY dim companion at only magnitude 11.4. Any other star isolated against a dark background sky would be a relatively easy dark sky/100x object for a 3-inch....but not when so overshadowed ("over-brightened?") by such a bright star so close. The faint star is a red dwarf, but due to its faintness the color cannot be seen. In a 6-inch (maybe with 220x) and perhaps a lot more easily with the 8-inch, look for this very faint star DUE SOUTH of the brighter star.

Object 3 - Another Very Tough Double - Zeta Aquilae
To give you an idea on just how difficult it will be to see this star, note that it was DISCOVERED in the 26 inch refractor at the U.S. Naval Observatory and had gone unseen before. However, that being said, the very faint companion star to Zeta (magnitude 3.0) can theoretically be seen even in the 3-inch; but that is "theoretical." It cannot be

seen at its 5" arc distance from the bright star even with the 6-inch; there is only a thread of hope with the larger 8" scope as well, but I have glimpsed this star on repeated occasions when the seeing is very steady and using powers in excess of 250x with 8" scopes of good quality. The faint star can be found (?) exactly northeast of the brighter star and VERY close; if you have a crosshair eyepiece, try blocking out the glare of Zeta with a cross hair and then look for the star. This trick works WONDERS in difficult double star observing!

Object 4 - A Very Nice Double for the 4-inch and Larger Scopes - 23 Aquilae
You won't find this one in your sky Library, so dial in the coordinates and try to find this magnitude 5.9 star and its 9.5 companion. Because of the star-rich region of Aquila, this star will be a bit hard to find, but once located, medium-high power (about 120x) in a 4-inch and larger scopes will reveal the fainter star almost DUE EAST of the brighter component, separated by only 3" arc, making it not an easy target for the small scope, but certainly should be resolvable. It is a very nice contrast in brightness and worth the effort to find.

Object 5 - Pi Aquilae - Tough Test for the 4-inch!
Good Double Star This pair, magnitudes 6 & 7 are a great test for a good 4-inch, at a separation of only 1.4" arc; in some cases it MAY be a good test as well for the 6-inch. Try it out and see if YOU can split it! Look for the stars in a NE-SW orientation and VERY close and almost equal magnitudes. If you do not split it at first, make sure you have the correct star, as this is just below naked-eye visibility. Then go up to about 100x....if it still does

not "open up" try 150x to 200x depending on the steadiness of your night air. That should do it!

Object 6 - Eta Aquilae - Very Nice Naked Eye Variable Star! Small telescope ideal object!

This is a very nice Cepheid-type variable star (see ASO GUIDES/Observational/*Observing Variable Stars* on the ASO website) that likely is more accurate than your wristwatch at keeping time. Really. Does your Timex keep on tickin' to the tune of 7.17622 day accuracy? I think not. The Delta Cephei stars are pulsators that brighten as they expand their diameters (like a balloon getting larger) and dimmer as they shrink. However, unlike MOST Cepheid variables, Eta Aquilae exhibits a very interesting deviation as can be clearly noted in the light curve below:

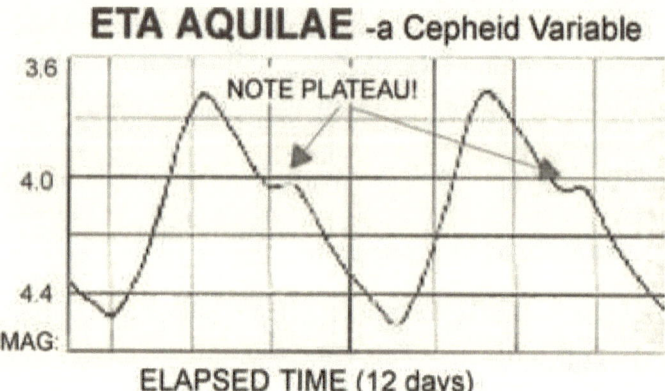

ETA AQUILAE -a Cepheid Variable

ELAPSED TIME (12 days)

Note the small "plateau" about one-half way as the star is diminishing in light. It last for nine hours and occurs with every dimming of the star; this is not a normal occurrence for Cepheid variables. The best way to observe this star is either with your lowest power on the small APO refractor or with the naked eye/binoculars. Since the star is relatively bright, comparison stars will be some distance off, but are clearly identified using the fine comparison and finder charts from the American Association of Variable Star Observers (AAVSO). These wide field "a" finder/comparison charts can be downloaded, saved to file, resized and printed in chart form from your computer. Because of its minor brightness changes AND the rapid period (it diminishes in less than one day!) there is NOT a dedicated chart for Eta Aquilae, by linking to www.aavso.org you can access the many charts available for other nearby stars that will include this star in their widefield ("a") charts!

Object 7 - A Very Fine and Easy Variable to Follow! R Aquilae
This huge star is a pulsating "Mira"-type variable red giant star that may be about 600 light years distant (it is difficult to judge accurately distances of Long Period "Mira" variables). It has an nice magnitude variation of almost 7 magnitudes and is easily observable throughout its range from a 3" and larger telescopes. However, the small telescopes are the ideal telescopes of choice for observing such stars, which range from magnitude 6 to about 11, fully within the range of these telescopes throughout its cycle.

R Aquilae -a Mira Long Period Variable

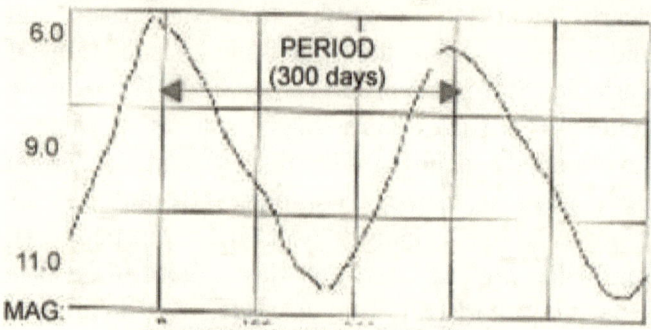

ELAPSED TIME (400 days)

Notice how predictable the light changes in this star are from the sample light curve above. The period (total number of days from brightest through dimmest and then back to brightest again) is actually DECREASING on R Aquilae. Once over 350 days, the period is now just under 300 days! Thus, you should observe this star at most only about every week or two and report your observations routinely to the AAVSO at the link provided above.

For a good comparison star chart for this interesting long period variable, link to: https://www.aavso.org/apps/vsp/ . Note for these charts, simply type in the NAME of the variable at top to generate your choice of chart. for a chart to make estimations of magnitude, and as the finder ("a") chart for both locating the star and for estimating with brighter stars (naked eye) when at its brightest.

Be sure and note the incredible RED color of this star throughout its cycles. As one of the coolest stars known (all of its friends are constantly saying:

"....man, you're the COOLEST star I've ever known...", R Aquilae is located in the "great rift", the dark star-obscuring cloud of the Milky Way, making its reddish color stand out even more vividly.

Object 8 - A Nice Small Galactic Cluster: NGC 6755
This is a relatively small (10' arc, about 1/3 the moon's diameter) but moderately bright (magnitude 8.3) cluster containing about 50 stars, most of which are too faint to individually see in all but the 5- and 8-inch scopes; even with those larger apertures, expect to see only 15 or 20 of the brightest members of this group on the darkest nights, with the remainder appear as a very dull background "glow." It is a very nice cluster, however, with some of the brighter stars visible in smaller telescopes show this - and the following cluster, ngc6709 - as a faint round glow in medium-high power. This is a VERY distant galactic cluster, about three times further than the only-a-bit-brighter ngc6709.

Object 9 - Another Small but Nice Galactic Cluster - NGC 6709
Located due north and a bit west of ngc6755 is another galactic cluster that is very similar in appearance, but much easier to see the fine array of stars at medium-high magnification. Total, there are about 40 stars visible in this cluster with the 8-inch, about 30 in the 6-inch and some 12 visible in the small APO refractor at high (120x) magnification. Note particularly the brighter "double" star in the photograph below taken with a 13" astrograph (courtesy Lowell Observatory) seen at the lower right of the cluster. These two stars are the brightest

members of the cluster, but several other brighter outer members are clearly visible. This is a nice object for very dark skies, one that is often overlooked in a very star-rich portion of the Milky Way.....you owe it to yourself to do some slow scanning through this area!

NGC 6709 - Galactic Cluster in Aquila

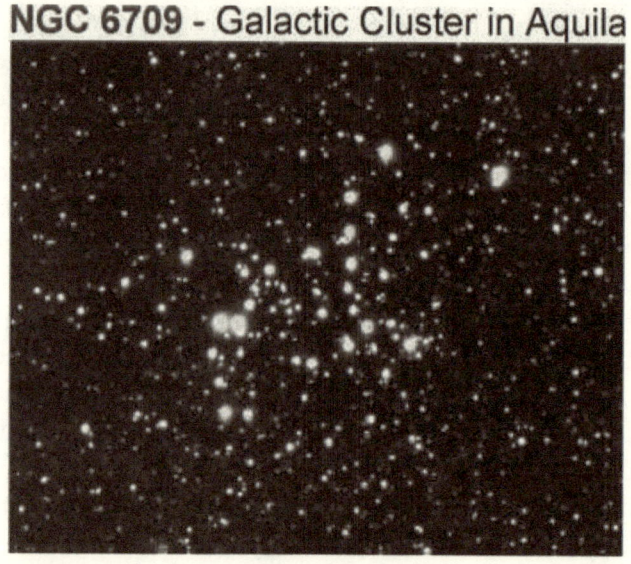

<u>Objects 10</u> - Very Nice **Dark Nebulae** - The BEST for Amateur Telescopes: B143 and B133 Two of E.E. Barnard's "dark nebulae," these light-blockers are excellent objects but require VERY dark skies and very low power, wide field views. The Milky Way is so incredibly rich with very faint stars and rich nebulosity in this region that dark nebulae on a very deep dark night really can stand out. B143 is located at R.A. 19h 38m; DEC +11 degrees 00m and is most definitely the easiest to find and see. Look for it only 1.5 degree WEST of gamma Aquilae; it is that easy to find. On the other

hand, B133 (R.A. 19h 05m; DEC -06 degrees 55m) is not as conspicuous and requires your GO TO function to locate. Look for it as a place where "stars ain't" as they say. In the rich Milky Way, both stand out clearly as black spots against the star field behind.

Dark nebulae is actually light-absorbing dust that rests amidst the spiral arms of the galaxy; indeed, were it not for such dust, the intense light and radiation from the center of our galaxy would make life difficult if not impossible throughout the galaxy! The dark dust is BETWEEN you and the stars blocked behind it that you merely cannot see.

Look in your lowest power with a fully dark-adapted eye (at least 15 minutes or more) at B143 for two lobes or "prongs" that can be clearly seen in most scopes. This double nature can even be seen in binoculars as the dark nebula is very large, about the size of the moon. B133 is about 1/3 that size and a bit more difficult to make out at first; moving the telescope very slightly back and forth across the field will assist in your eye adapting to such a low contrast object.

Object 11 - Another Excellent Variable Star! **R Scuti** (in Scutum)

This is another very fine classic variable star that is well worth keeping up with, and easily observable - if not the BEST - star for smaller telescopes, varying between magnitudes 5.8 and 7.8 in a period of only 140 days; thus it is a good idea to record this star's brightness about once per week. Actually, if you are in a hurry, you can easily make your magnitude estimates with some good 7 x 50 or 10 x 50 binoculars! However, you will be missing the incredibly stunning field of stars surrounding this area. It is a very rich region near the Milky Way arm.

R Scuti is a classic and easy-to-follow "semi-regular" RV Tauri type star that has an "overlying" period of some 140 days, but also exhibits considerable unpredictable and erratic behavior within that period.

Download and save to file this link from the AAVSO for R Scuti as your observing and locating chart for this nice variable: https://www.aavso.org/apps/vsp/ . Note for these charts, simply type in the NAME of the variable at top to generate your choice of chart.

Notice in the figure below the remarkable difference in the light curve for R Scuti than for our preceding Long Period and Cepheid variables. R Scuti is only MODERATELY regular as can clearly be seen! And that is what makes this star so remarkable exciting to observe! In addition to the regular large brightness fluctuations of not-even "regular" periodicity, there are innumerable minor outbursts and unexpected variations to make this a truly exciting star to keep up with.

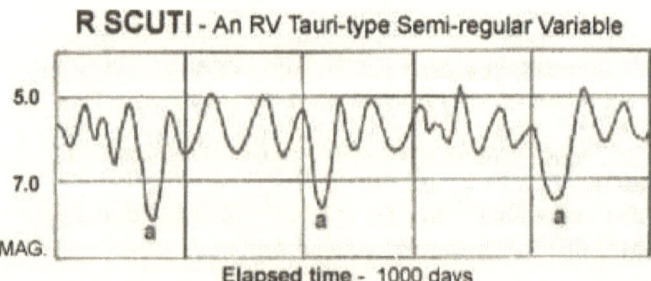

R SCUTI - An RV Tauri-type Semi-regular Variable

Elapsed time - 1000 days

Note the deep minimum periods (marked with the

red "a") that occur on a semi-regular basis as well as all the minor irregular fluctuations in between. The "primary period" is about 144 days, but the star is so irregular that much can happen in that interval. You should make an effort to monitor this star to assist in the studies that are underway on it and other stars like it. Send your observations once per month (observe about once weekly) to: www.aavso.org to have them included in the growing body of knowledge of the semi-regular variable star groups.

Objects 12 - the BEST galactic cluster of the summer - Messier 11 - "*The Wild Duck Cluster*"

Messier 11 is like Saturn when you see it for the first time in your telescope....you will never forget the beautiful view. This cluster almost resembles a sparse globular cluster but is technically a very RICH galactic cluster some 5000 light years away; one can only image what this remarkable object would look like if moved to the distance of our CLOSEST star clusters, such as the Pleiades, only 450 light years distant! Even at that great distance, the 8" scope can reveal about 250 stars down to magnitude 13.8 easily; only slightly less stars can be seen with the 6-inch; expect some 175 plus stars with3-inch and using medium high magnification with the small wide field scopes you should be rewarded with at least 100 or more beautiful stars....all about the same magnitude! Using medium magnification really reveals the beauty - and more stars - of this cluster. In reality this wonderful object contains an estimated 850 to 1000 stars down to magnitude 18.

The first thing you will notice about M-11 is the interesting yellowish bright star right in the middle of the cluster, obviously MUCH brighter (magnitude 8.0) than the stars of the cluster. There is a reason for that. This particular star is a FOREGROUND object....it is NOT part of the cluster, but located somewhere between M-11 and Earth. It is this star, on a very dark night, that provides a near-3-D look to Messier 11 and its surrounding rich stellar neighborhood. It was the famous observer *"Admiral Smythe"* who gave this object its aviatric name "Wild Duck Cluster," when he remarked in his observing notes that this remarkable object appeared to him as a "....*flight of wild ducks*" in the sky.

Object 13 - Galactic Cluster Messier 26 (a mere kid-brother to fabulous M-11)
Messier 11 is wonderful, and is a tough act to follow, and is only magnitude 9.1 and about 10' arc across; thus it appears as a disappointment if you have just looked at Messier 11! With a good dark night and the 6-inch about 25 brighter stars can be seen out of maybe a total of 125. About a dozen are visible in the 3-inch and the brightest 9 are clearly visible in small telescopes under medium high power. Messier 26 is truly smaller and contains less stars than M-11, as both are almost the same distance away.
Object 14 - A Small, but Fine Globular Cluster - NGC 6712
A lone globular cluster - but one that is worth checking out - is located in this "GO TO" TOUR, this being in Scutum. At magnitude 8.3, it is a fairly bright and easily-found object in the 4-inch and larger scopes; it appears as a very tiny and faint "fuzz ball" to the small telescope. Medium power (about 30x per inch) is required to really get a good

look at this object, but even with the 8" no resolution is possible, this object being very distant. On a very dark night and about 220x, a good 6-inch should see some "granulation" of the small disk (only 2.5' arc) of light around its edges, these being the red giant stars that are so faint as to be just out of reach of our range of telescopes.

WANDERING ABOUT....YOUR NEW "USER OBJECT" IN AQUILA

Since the two dark nebulae (Barnard's B143 and B133) are perhaps the easiest and best such objects for amateur telescopes, AND you may want to try your hand at some piggyback astrophotography (see my "Piggyback Astrophotography" guide under the GUIDES tab on this ASO website). LET'S LOAD both these dark nebulae INTO OUR USER OBJECTS! Take your Autostar and key in the coordinates for each as given above.

On AutoStar, go to: "Select/Object [enter]...." scroll down to "User Object" [enter]. Now enter the coordinates given above for "B143", using the number keys on AutoStar. After entering the coordinates and pressing "Enter" yet again, scroll down one and you can list the magnitude of the object as "0"[Enter]. Now go back and mode back to "USER OBJECT / Add...." and enter the coordinates and information for B133 as well.

Be sure and check out SAGITTARIUS, the "Archer," tour in Volume II *Constellations*; this constellation, just south of Aquila, is the home of the most fantastic view of star clouds this galaxy has to offer from Earth! We will explore the Omega nebula, the Trifid, the great Sagittarius star cloud, and peer right into the galactic center and its impacted masses of stars and dark nebula. In the meantime, hone up on your piggyback astrophotography skills with your telescope....this is one area of the sky where a simple digital camera set to "bulb" (B) and a five minute exposure can reveal the true wonders of our Milky Way galaxy! In this wonderful majestic soaring celestial Eagle we will examine the rich and thick clouds of stars that comprise the most dense portion of our Milky Way galaxy as we move every-so-closer to the galactic center in the famed archer Sagittarius.

Good Observing and explorations of this wonderful world of deep space!

*   *   *

Chapter Four
ARIES
**

With a Bonus Listing of Beautiful and Challenging Double and Multiple Stars!

On the Horns of A Mighty Ram Leading the Zodiacal Herd

This is the fourth Constellation Guide, "GO TO ARIES" of the *Constellation* series for all computerized telescope users.

An important "astrological" constellation, Aries represents the *FIRST SIGN* of the *ZODIAC* (see the astrological circle shown below), a heavenly Ram that is nearly buried among brighter constellations.

Because of annual precession (the relative positions of celestial objects as we see them changing from Earth due our annual motion of the solar system as the sun moves through the **Milky Way** as well as the Earth's changing orientation on its axis), Aries is now located beginning at astronomical hour angle roughly 1 hour and 45 minutes, and extends to about 3 hours 30 minutes of Right Ascension. It is so small that it seems almost intentionally "tucked" exactly between declination lines of +10 and +30 degrees.

Although the "number one" zodiacal constellation, this is a very minor area of the sky in terms of bright showcase objects, Messier objects and bright stars. Indeed, it is not even FIRST in the alphabet, being the sixth constellation in that order. Like other constellations that have been discussed in these "GO TO" tours of our constellations, Aries is conspicuously absent of: any star brighter than magnitude 2.0; any Messier Object; globular clusters, galactic clusters, planetary nebulae, diffuse nebulae.

The only thing really going for this "RAM" of a constellation is that the *ECLIPTIC* - the imaginary band in which the sun, moon and planets appear to traverse as we watch them from Earth - pass from the southwest corner to nearly the middle of the eastern side of the constellation.

Indeed, there are ONLY seven NGC objects - all galaxies - that are even distinct enough to view in common (yet large) amateur telescopes. There are two (2) diffuse nebulae....VERY small, but bright enough to be seen in a telescope....that are not even numbered on the NGC catalog: one at 03h 23m / + 30 46 and the other at 03h 25m / + 29 39 (9.0 and 9.3 magnitudes and 4' and 11' arc in diameter, respectively). These are very difficult to spot, in spite of their sizes.

Clay Sherrod's
CONSTELLATION GUIDES:
"Go To" ARIES the "Ram"
The First "Sign" of the Zodiac·

A finder chart for locating many of the GO TO objects in the constellation of Aries; if using a computer planetarium program, you are encouraged to plot the objects on your screen for higher resolution than this chart provides.

*   *   *

There are, however, some very beautiful double stars in Aries and those are the focus of this GO TO guide to this constellation for the most part.

It is not just in modern telescopic times that Aries has been a largely ignored chunk of sky. Even the star cataloger **Bode** rather "skipped over" the mighty RAM when he compiled his beautifully-illustrated 19th century star atlas. Look at the sample below, which shows the grandeur of *Perseus* and *Andromeda*....even the lowly TRIANGLE of *Triangulum*. But WHERE is *Aries* the Ram? Not shown. But Bode did, nonetheless, bother to at least "label" the constellation by name if you look hard enough. Nonteless, the far superior art of Johannes Hevelius did very distinctly honor the heavenly goat.

Aries (pronounced "AIR-eeez") is rather uninteresting naked eye constellation of fall skies. Like Andromeda, it is in that "quiet zone" that follows the spectacular summer Milky Way clouds and wealth of rich deep sky objects.....yet precedes the equally rich skies of winter, harboring the beautiful and bright constellations of Orion, Taurus, Cassiopeia and Canis Major. To me, Aries has always looked a bit like a ".....crooked dog's hind leg," which probably indicates something significant about my mental health.

Aries rises NORTHEAST for mid-northern latitudes about 9 p.m. local time on about August 16. High in the north, it takes a while to reach the meridian, or highest point in the sky, culminating at 4:15 a.m. on the following morning! However, midnight culmination - when it is on the meridian at midnight each year - occurs always on about October 20.

As with every "GO TO" tour guide, each GO TO object in Aries is discussed for your telescope regarding the type of conditions necessary for you to view it optimally for discern the very faintest details.........magnifications and aperture necessary for most objects, and much, much more. This is YOUR complete guide to get you on your way to exploring the best (and few!) objects in this small constellation. The following listing of "BEST" objects contains the finest or most interesting from my own observing experience and preference.

Use the attached star chart and the following Guide as an excellent reference for your next star party itinerary, or a beginning for further study into the thousands of objects visible in this part of the sky. Truly these extensive Constellation Study Guides will most definitely put your computer program to work for you in the most efficient and enjoyable way possible! As a matter of fact, MANY sky program users are now programming their own "Tours" based on these guides, using each constellation as a separate GO TO Tour for the sky library that can be added in or deleted through the main edit screen on your PC or MAC computer.

We hope you enjoy these comprehensive guides to touring the constellations via your computer-driven telescope. Each new installment is complete with diagrams, charts and illustrations that you will find nowhere else. Please let us hear YOUR feedback and your observations of each and every constellation after YOU have toured its vast reaches of our skies!

YOUR ARIES CONCISE DIRECTORY OF INTERESTING OBJECTS –

As mentioned, there is a void of interesting deep sky objects in Aries. So if you are a certified "galaxy hunter" or diffuse nebula groupie, this is not the constellation for you. However, if you enjoy beautiful and challenging double/multiple star observing, there are some wonderful targets for you in the Ram. For a full discussion on double star observing and their "**Position Angles**" refer to my brief overview in the "GO TO" tour guide for **Lacerta**: guides: *Constellation Lacerta* found in Volume II of *Constellations*.

I have chosen the most interesting 14 targets in this ARIES "GO TO" TOUR; as with all GUIDES, all objects listed below will be visible in most telescopes (some naked eye) from scopes in size from 3-8 inches; of course larger apertures may "show" an object a bit closer and "better," but frequently a wide field and low power view is more desirable than aperture for FINDING the objects initially. Indeed, I strongly encourage you first FIND the target object, or its approximate location through your GO TO function with your lowest power and then - once IDENTIFIED positively - move up slowly in steps with magnification if necessary. Remember, not all objects "like" magnification. Sometimes better "field of view" (such as the wonderful wide fields provided by smaller telescopes) is desired over light gathering (like an 8-inch) and magnification. Note that your sky program may NOT have every object listed on every constellation GO TO tour....this is intentional. You can access some of the most interesting objects of the sky directly from their coordinates. It is quite simple as you merely enter these coordinates as follows in the 10-step process:

The constellation tour Star Chart will get you started on your journey for this constellation.

Following is the concise object list for your "GO TO" TOUR of this constellation; you may wish to find many of the objects from the AutoStar or other sky Library. For example, you can easily go the pretty 7th magnitude open cluster, Messier 52, if you pull up "Object/Deep Sky/Messier Object/..then type in '52'...." and then press "Enter", followed by "GO TO" to access this rich cluster. On the other hand, if you want to experiment and become a "better AutoStar user" try entering the exact R.A. and DEC coordinates of that object as given below after holding down the MODE key. You will find the accuracy of entered GO TO's to be somewhat less than those stored in AutoStar, but the capability of acquiring unlisted objects is fantastic!

You will access your FIRST GOTO target - (usually the brightest star in each constellation) - via the command "SETUP / OBJECT / STAR / NAMED....and scroll to **"HAMAL"**, then press "Enter" and subsequently "GO TO" to move your this bright star.

You may also access the constellation by: SETUP/OBJECT/CONSTELLATION/Aries.....Enter....GO TO, which will subsequently take you to the brighter star Hamal, near the middle of the "crooked hind leg of the dog."

OBJECT 1: brighter star HAMAL (alpha Arietis) - R.A. 02h 04' / DEC + 23 14 - Magnitude: 2.0
OBJECT 2: bright double star - MESARTHIM (gamma Arietis) - R.A. 01h 51' / DEC + 19 03 - Mags: 4.5 & 4.5! Super!
OBJECT 3: nice double - 30 Arietis - R.A. 02h 34' / DEC + 24 26 - Mags: 6 & 7 - VERY nice for all scopes - WIDE!
OBJECT 4: runaway star - 53 Arietis - R.A. 03h 05' / DEC + 17 41 - The "runaway" star from the Orion Nebula!
OBJECT 5: Struve 326 - R.A. 02h 53' / DEC + 26 40 - Mags: 7.5 & 9.5 - wonderful double, enough space for all!
OBJECT 6: elliptical galaxy - ngc697 - R.A. 01h 49' / DEC + 22 06 - Mag. 12.2 - VERY faint, 6-inch & up.

OBJECT 7: faint spiral galaxy - ngc772 - R.A. 01h 57' / DEC + 18 46 - Mag. 10.8 - Visible as speck in 3-inch

OBJECT 8: another faint spiral - ngc972 - R.A. 02h 31' / DEC + 29 06 - Mag: 12.5 - this one for larger scopes only.

OBJECT 9: irregular galaxy - ngc1156 - R.A. 02h 57' / DEC + 25 03 - Mag: 11.8 - faint, but visible in a 6-inch +

OBJECT 10: very nice variable star - R Arietis - R.A. 02h 13' / DEC + 24 50 - Mag: 7.3 to 13.8 in only 187 days!!

*** BONUS: ***

Your Aries List of <u>VERY NICE DOUBLE STARS</u>!! Here is a list of doubles in Aries, appearing in abbreviated form. On each of the following stars, you will find the following: A = **star designation** / B = R.A. / C = DEC / D = magnitudes of components / E = separation in arc seconds (") / F = minimum telescope in inches required for resolution under BEST conditions / followed by any comments if needed! (these stars are in order of Right Ascension and DO NOT include those for the TOUR):

1. **#1 Arietis** - (B = 01 47) (C=+22 02) (D=6 & 7.5) (E=2.7") (F = 3) Beautiful gold and blue!!

2. **Struve 175** - (B = 01 48) (C=+20 52) (D=6 & 8.3) (E=23") (F = 3) Tough to find, but worth the effort

3. **lambda Arietis** - (B = 01 55) (C=+23 20) (D=5.2 & 7.6) (E=37") (F = 3) Nice and easy!

4. **#11 Arietis** - (B = 02 04) (C=+25 28) (D=6.4 & 12) (E=1.6") (F = 5) Quite a test!!

5. **Struve 226** - (B = 02 09) (C=+23 44) (D=8.1 & 10) (E=1.8") (F = 3) Tough test.

6. **Struve 43** - (B = 02 38) (C=+26 25) (D=7 & 8.6) (E=1.1") (F = 5) Very tough even in 8-inch.

7. **mu Arietis** - (B = 02 40) (C=+19 48) (D=6 & 12) (E=19") (F = 6) Doubtful except with 8"

8. **pi Arietis** (triple) (B = 02 47) (C=+17 15) (D=5, 8.5, 10) (E=3.2 & 25") (F = 3) really neat!!

9. **epsilon Arietis** - (B = 02 56) (C=+21 08) (D=6 & 6) (E=1.4") (F = 4) great! Aligned nearly N-S!!

10. **Struve 366** (triple) (B = 03 11) (C=+22 46) (D=7, 10, 10.5) (E=47, 1.7") (F = 4) faint star is double.

A VISUAL GUIDE TO OUR DEEP SKY OBJECTS IN ARIES

<u>Object 1</u> - Our "Starting" Bright Star - "HAMAL" (alpha Arietis)
This star name demonstrates the common thread among ancient civilizations. When you consider so many celestial objects - the "lion" of Leo is a good example - that share the same connotation among the Chinese, the Egyptians, Arabs and other ancient historians, you must wonder how the "word spread" among these different corners of the world to hold such a similar attachment to mythology. The Arabic name for the brightest ("alpha") star in Aries is

"Hamal", meaning the *"sheep's head,"* a slight deviation from the later Greek reference for this star as a mighty Ram. Nonetheless, all seem to at least be swimming in the same waters regarding the overall "family" of animals! With an actual light/energy output some 70 times greater than our own sun, Hamal is a relatively close star, only 75 light years distant. Look for the distinctive "red-orange" color of this very late spectral type star.

Object 2 - "MESARTHIM" - (gamma Arietis) - One of the Most Popular Double Stars
This star has special meaning to old Doc Clay....it was the FIRST double star that I turned my brand new Unitron 4" refractor onto after it arrived on a brisk October night atop Cherry Hill in 1966. Prior to that night I had spent three years learning the entire sky and all the objects within the constellations with a pair of old 7 x 50 marine binoculars from my Grandfather's Chris Craft "marine vessel" as he coined it.

This is an absolutely beautiful double star and one very easy for all telescopes in all sizes. The components are too bright an close to miss! Both are almost identical in brightness (4.7) and color (brilliant white). Even at low powers the stars are easy to split because of the wide 7.8" arc separation; the color is best appreciated at about 15x per inch aperture of your telescope, which also provides a very pretty star field around this pair as well. The slightly fainter star is in Position Angle 48 degrees, or almost due NE from the brighter star.

Object 3 - 30 Arietis - A very nice and easy double star for all scopes – *see chart, next page.*

At a distance of only 190 light years, this is a wonderful double star for common telescopes. The brighter star is distinctly yellow in medium magnifications at magnitude 6.6, while there is a noticeable "blue" tinge to the fainter (magnitude 7.4) star which is nearly due WEST (Position Angle 274 degrees) of the brighter one. The separation of nearly 40" arc makes this a prime candidate for smaller instruments!

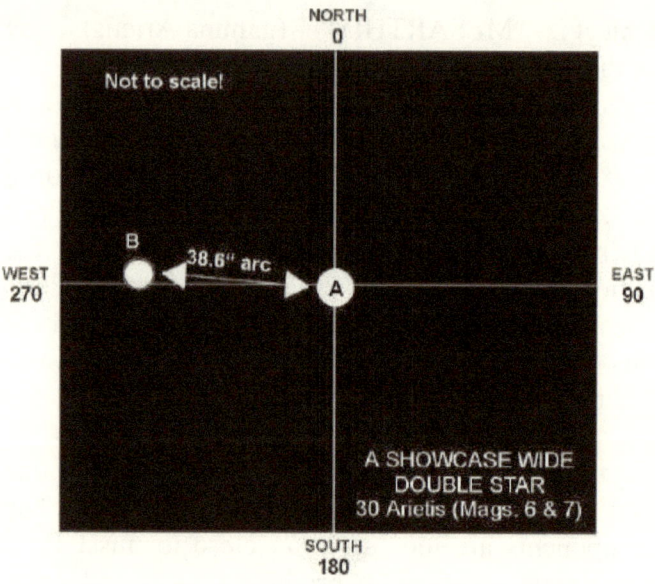

Object 4 - **"The Runaway Star"** - 53 Arietis.
One of three such stars that appear - if you back-trace their pathways from their current large proper motions through the sky - to have been "ejected" from the region of the Orion Nebula! Of the hundreds of known stars that are associated with the nebula, and likely FORMED from it, these three somehow "got away" and exhibit extremely high velocities in space, with 53 Arietis moving at an incredible clip of 35 miles per second....thirty five

times faster than the X-15 Rocket Plane of the U.S. Air Force!

If we backtrack from where these objects are today....and factor in the speed at which they are moving....then astronomers can calculate that they "left" or were pushed out of the Orion Nebula about five (5) million years ago - just yesterday in the cosmic scheme of things.

53 Arietis is magnitude 6.0, easily visible in all of our telescopes provided you know WHICH star you are needed to focus in on. Use my star chart below (adapted from the A.A.V.S.O. variable star charts) to help acquire the actual field of view of this star. You should notice that - just like the very hot, young stars of the Orion nebula - this star appears brilliant white. The other two "runaway stars" are AE Aquarii and Mu Columbae (southern skies), and all three are erratic variable stars.

YOUR FINDERCHART FOR THE "RUNAWAY STAR:
53 Arietis
Chart Courtesy Arkansas Sky Observatory

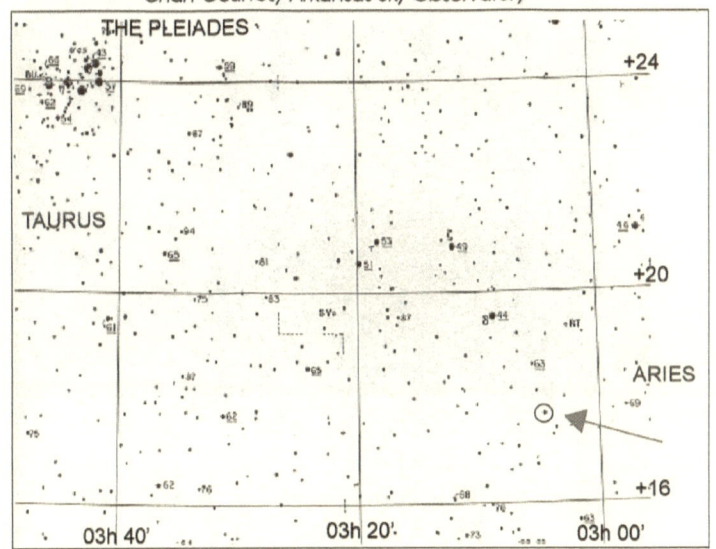

Object 5 - Struve 326 - A Fine Red Dwarf Double Star

Both of these stars are smaller than our own sun and very close to our neighborhood, at 55 light years distant. The colors of the stars (crimson red) are very distinctive in a small telescope with powers of 100x or greater. This should be a relatively easy split for even small telescopes with a separation of about 6.5" arc. The brighter star is magnitude 7.5 while the fainter one comes in at a rather dime 9.5, just west of due south from the primary star (Position Angle 219 degrees). Although the star should be rather easy in a 3-inch, the chart below will assist in locating the fainter star relative to its fairly bright partner.

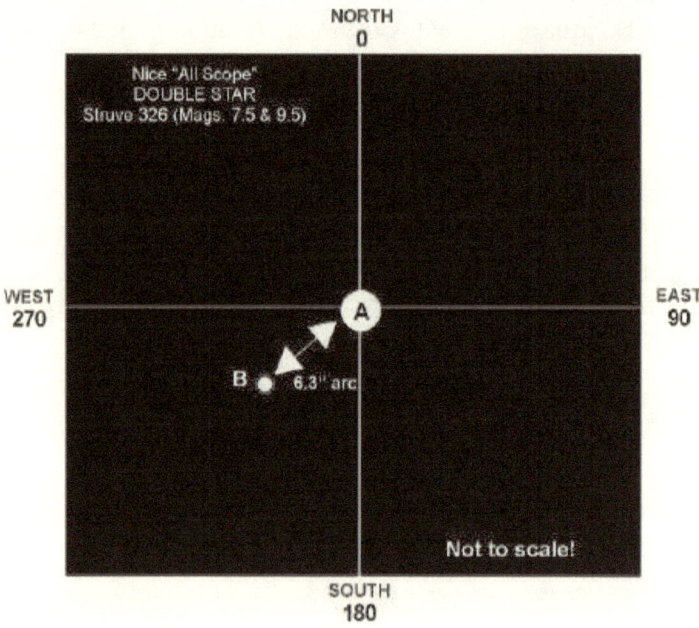

Object 6 - A very faint elliptical galaxy - NGC 697

This will be a challenge for a 6-inch scope and likely NOT seen except as perhaps a very faint starlike object some 2' arc across, like an out-of-focus star. The 8-inch will show it a bit better and clearly differentiate its nature from that of a star. This elliptical galaxy is oblong (2' x 1' arc) and only magnitude 11.8, thus is right at the "extended object" magnitude limit for a six-inch scope, and close for the 8" scope. Forget this one with smaller instruments.

Object 7 - A faint spiral galaxy - NGC 772
Here is another faint galaxy for the strong at heart. Its magnitude 10.8 is deceiving....this galaxy is only 5 x 3' arc across and very difficult to find. It should be relatively easy if your skies are very dark with a 5" scope, but do not expect much detail on this object at all. This is what we call "grabbing at straws" for GO TO objects....remember we are in a zone of "nothing" in the way of deep sky objects here. In an 8-inch, there may be a HINT of a very large "lobster claw" spiral arm to the northwest of center....use about 225x to get the best chance of seeing this appendage.

Object 8 - Another faint spiral galaxy - NGC 972 - Blotchy at Best!

You can be the judge on this one, based on the photograph shown above from the Palomar 200" telescope. Is this a true "spiral galaxy" or perhaps an "irregular?" I can clearly see the indication of spiral structure, but this is truly a galaxy in turmoil. This object will ONLY be visible in 8-inch and larger telescopes, at magnitude 12.5. It will appear extremely tiny, like a fuzzy star right at the threshold of visibility. Use very high power (as much as the object will hold before becoming too faint to see) and you might see a bit of mottling within this tiny object.

Object 9 - An irregular galaxy for the books - NGC 1156
Another VERY faint galaxy, but an interesting irregular one that shows much chaotic structure as can be seen in the Palomar Observatory photograph

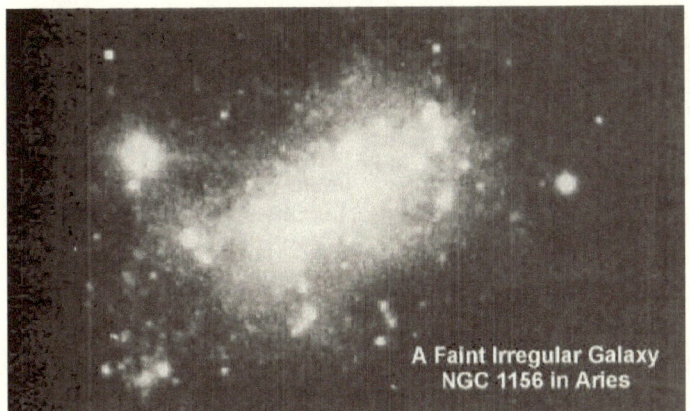

A Faint Irregular Galaxy
NGC 1156 in Aries

above. This galaxy, like all in Aries, is just on the verge of visibility in most 8 and 10 inch scopes and users of smaller instruments need not waste their time....find a good double star to play with. There is really nothing interesting to see with this object, even in the largest of all amateur telescopes.

Object 10 - A Very Nice Variable Star - R Arietis R Arietis is classified as a long-period variable star, but it has a wonderfully "short" period in spite of it - only 187 days! During that period, the star will brighten to magnitude 7.3 or thereabouts and subsequently dim again to about magnitude 13.8 and eventually back to brightest again! This cycle can be completely monitored with a 3-inch scope and larger scopes provided that the user locates and identifies correctly the right star. This is where the American Association of Variable Star Observers (AAVSO) star charts come in. Free for the downloading, you may download to file the FINDER "a" wide field chart at:
https://www.aavso.org/apps/vsp/ . Note for these charts, simply type in the NAME of the variable at top to generate your choice of chart which will give nice reference magnitude comparison stars for when

R Arietis is at its brightest. As it begins to fade fainter and fainter, move up in magnification to allocate even fainter stars through the AAVSO charts. The "g" chart shows stars to less than magnitude 14 (far beyond the reach of nearly all scopes) for comparing the star as it plummets in brightness. Note that BOTH of these nice charts cann be "REVERSED" charts, meaning that they have been constructed to match the view of a Maksutov or Schmidt Cassegrain telescope with the diagonal mirror in place (north at top and EAST to the right).

Merely download each chart and assign a file name; bring up that file and resize the image (it will be quite large) to fit the page and "save." Then bring that re-sized file back up and print it! You now have two wonderful star charts to devote many nights observing this wonderful star. I recommend observing the star about once per week but really not more often, as daily observations (as you might do with an irregular or eruptive star) are overkill with little or no change seen from one day to the next. Log your estimates, plot the on your own graph paper and see what YOU come out with! And by all means e-mail your results on a monthly basis to: www.aavso.org so that your observations can contribute the scientific body of knowledge for this star!

WANDERING ABOUT....YOUR NEW "USER OBJECT" IN ARIES

Red dwarfs, white dwarfs, neutron stars, black holes, X-ray sources, red giants....sounds like a ZOO! So why not add a "runaway object" to your

"GO TO" guide of user objects in your Autostar or PC sky library?! Not only is 53 Arietis one of only three such known stars to be "expelled" from the Orion nebula and flying through space, but it also is a moderate variable star as well! It varies from magnitude 6.1 to 6.3 (not really enough to monitor with the human eye) in addition to its antics as it moves freely through space!

On your sky program go to: "Select/Object [enter]...." scroll down to "User Object" [enter]. Now enter the coordinates given above for "runaway", using the number keys on AutoStar or similar. After entering the coordinates and pressing "Enter" yet again, scroll down one and you can list the magnitude of the object as "6"[Enter].

Now you have yet another unique and oft-overlooked deep sky object for conversation-starters and crowd-stoppers at your next star party or club outing.

*   *   *

Be sure to check out the constellation CETUS, yet another constellation from the watery depths of the celestial rivers and oceans! This great Sea Monster / Whale (depending on which dramatic school of thought you follow) is one of *Ptolemy's* ORIGINAL 48 constellations from 2 A.D. from the 1022 brightest stars of the sky. It has remained virtually unchanged since his inception over 2000 years ago.

Good Observing and may the stars serve as your sentries as you explore the frontiers of space!

*   *   *

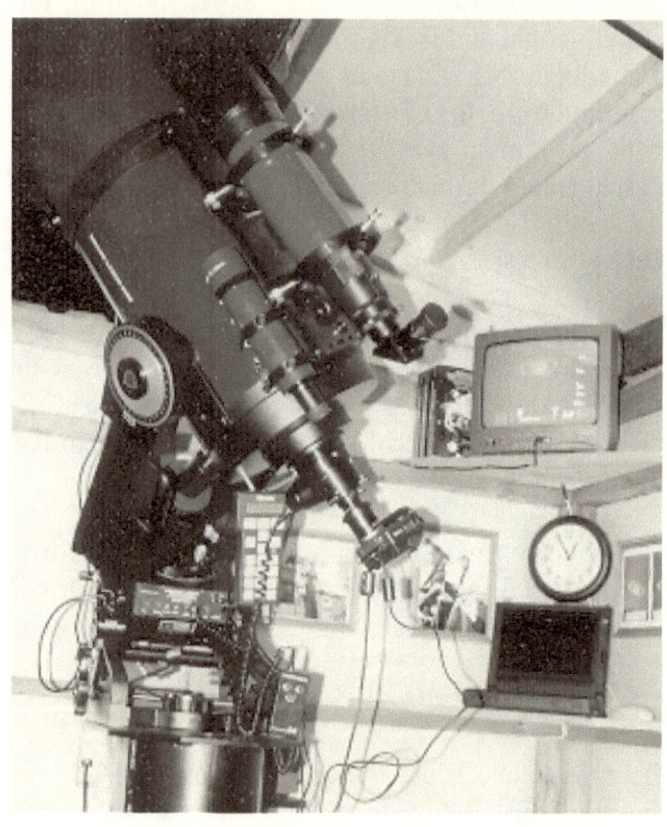

One of Arkansas Sky Observatories'
Robotic Telescopes, this one located at MPC H41
Utilizing advanced computer GO TO, but also
capable of on site operation using the Meade hand
control.

As you can see – expect some degree of wires for
most computerized applications!

Chapter Five

AURIGA
....but, say, isn't that a baby GOAT on his shoulder?

The magnificently large constellation of AURIGA is next *Constellation* tour, " for the modern GO TO computerized telescope. As we enter the realm of the celestial charioteer, note that we are moving deeper into the "Orion Arm" of our Milky Way galaxy; from our vantage point of space - our sun as but one star in a spiral arm of our galaxy - we are peering OUT of the galaxy to intergalactic space, looking across the most distant Orion Arm of stars.

Hence, you can expect the number of stars and galactic clusters within this more distant spiral arm to increase dramatically and there is no better place to witness this density than within the boundaries of Auriga.

You will note that this large constellation forms a five-sided "pentagram" of sorts (see my star chart below); however, this is actually not the case from a traditional standpoint, even though most star maps (including this one!) also include Beta Tauri - Al Nath - in the five-sided asterism. Al Nath (or simply Alnath) is located precisely on the south border of Auriga and the north border of Taurus and hence it is somewhat of a "shared" star between the two, but the actual Bayer assignment goes to Taurus on this one.

In addition to familiar Taurus, the Bull, to the south, Auriga is bordered at the north by Camelopardalis,

Perseus to its west, Lynx to the east, and Gemini to the southeast.

The *Aurigid* meteor shower is generally observable between January 31 and February 23, a fairly scant meteor shower that is better known for its bright fireballs than a large number of hourly meteors. It does not have a clearly distinguishable radiant, but rather the meteors appear to emanate from the direction of the entire group of stars that form Auriga. From August 25th until September 6th, the *Alpha Aurigid* (so named for the proximity of the shower radiant to the bright "alpha" star Capella) shower is active. Although the hourly maximum is about 9 meteors, outbursts of up to 30 meteors per hour were observed in 1935 and 1986. Another meteor shower - the *Delta Aurigids* - has an annual showing in Auriga. These may be observed between September 22 and October 23 each year with the maximum of this shower on around October 6 to October 15.

There are monthly Sky Calendar listings for each month of every year on the ASO website: www.arksky.org which provides dates and details of all yearly meteor showers for naked eye enjoyment.

Within the borders of Auriga are at least 10 brighter galactic stars clusters of interest to amateur astronomers, all of which are visible in the widest range of telescope sizes, and certainly most even in binoculars. Three of those are designated as Messier Objects, although several others could just as easily be so assigned.

Clay Sherrod's
CONSTELLATION GUIDES:
"Go To" AURIGA: the "Charioteer"
As bright CAPELLA heralds the Impending Winter

CAMELOPARDALIS

LYNX

[4]

[11] O 2149

[8]

[1] Capella

PERSEUS

Mekalinan

AURIGA

Sadatoni

[3]

[6]

[2]

O M-38

GEMINI

[9]
M-36

[10]
O
M-37

[7] Hassaleh

[5]

Nath
(TAURUS)

Bulletins of the

ARKANSAS

Sky

OBSERVATORY

Lat.
+35 05

Lon.
-92 30

THE ECLIPTIC

M-1

TAURUS

A finder chart for locating many of the GO TO objects
in the constellation of Auriga; if using a computer
planetarium program, you are encouraged to plot the
objects on your screen for higher resolution than this
chart provides.

*   *   *

Auriga (pronounced "awe-REE-ga") mythologically has several designations, and there is much uncertainty as to specifically which original story is accurate (well, actually is ANY myth really accurate?). In some interpretations, Auriga represents the god of the seas - *POSEIDON* (or perhaps "Neptune") - but he is STILL a "charioteer" in such variations nonetheless. Only in this case his *Chariot of Conch* is being drawn by very large (and I must say frightful-looking!) "sea horses."

The interesting aspect of the entire confusion "charioteer" story is that the constellation has always depicted Auriga without ANY chariot whatsoever, although the reins of such an unseen vehicle are clearly typically shown in his right hand. Another confusing aspect is this association he has with GOATS....yes, GOATS. Here is either a mighty and brave charioteer of Greek hero status OR a Greek god of the seas, and he has this fascination with GOATS! On his left shoulder is always depicted a small goat, while in his left arm he cradles two kid goats tightly as he "chariots-away."

So....yet another Greco-Roman myth has a slight angle on the entire chariot/goat combination, this portraying our Auriga NOT as a mighty wheeled warrior nor god, but the actual INVENTOR of the chariot - *ERECTHONIUS*, the lame son of the mythological notables **Vulcan** and **Minerva**. A lover of gentle animals and handicapped, it is told that **Erecthonius** actually invented the idea of the wheeled chariot as merely a way to get around, as he could not walk. So...in this case, it is not an instrument of war or fighting...it is in essence a wheelchair for a kind and gentle young man as he traveled near and far to care for the gentle animals throughout the lands of milk and honey.....

To the naked eye, the constellation of Auriga forms a conspicuous geometrical outline, with the bright yellow star *Capella* marking the advent of fall. At latitude 35 degrees north, Auriga is one of the few constellations that actual pass directly overhead - through the zenith. Capella culminates (reaches nearly directly overhead at midnight) each year during the first week in December. Although not circumpolar like Cassiopeia, Auriga is nearly so, with Capella rising very far in the north-northeast at 8 p.m. local time on October 1 and reaching high overhead not only until a full eight hours later - 4 a.m. the following morning.

In the diagram below, note the familiar asterism **"the Kids"** of Auriga, suggesting three baby goats that the charioteer is carrying cradled in his left arm. The first of these naked eye stars is ZETA Auriga, or *"Sadatoni"*, a variable star of about magnitude 3.8 but suspected of being slightly variable in brightness; next in the baby goat nursery and just slightly east is ETA Auriga (sometimes called *"Hoedus II*, with Zeta paired as *"Hoedus I"*), magnitude 3.3. The most "famous" of the "Kid Stars" is EPSILON, the northernmost of the three. It is a VERY unusual eclipsing binary star (hence slightly variable from magnitude 3.0 to 3.8, with the dimmest point lasting well over one year! The largest component of Epsilon Auriga is perhaps nearly 200 times larger than our own sun....the other star, also an even-larger supergiant star is likely 3,000 times LARGER than the sun!! That is an incredible three billion miles across! Yet it is also perhaps the COOLEST (in terms of temperature, not because of a fashion statement) of all stars known.

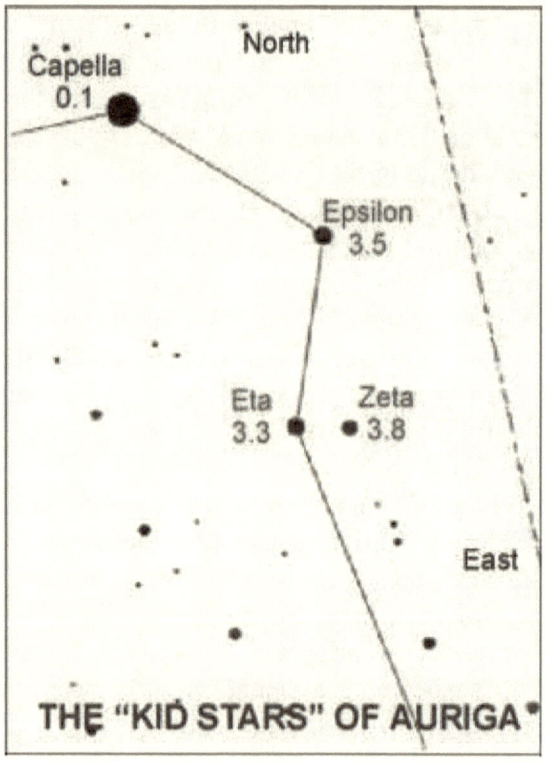

THE "KID STARS" OF AURIGA

For the full listing of all brighter stars (to naked eye limit) which includes: 1) Bayer; 2) Flamsteed; 3) SAO #; 4) R.A. & DEC; 5) brightness; 6) most notable double and multiple stars; 7) variable star designations; I encourage you to visit the wonderful Constellation reference site
http://www.deepskywatch.com/deepsky-guide.html
as well as all other constellation Star Tables. This is a great reference to print and put into a binder for cross referencing star designations, locating double

and variable stars, and current coordinates in the sky.

As with every "GO TO" tour guide, each GO TO object in Auriga is discussed for your telescope regarding the type of conditions necessary for you to view it optimally for discern the very faintest details.........magnifications and aperture necessary for most objects, and much, much more. This is your complete guide to get you on your way to exploring the best (and few!) objects in these two constellations. The following listing of "BEST" objects contains the finest or most interesting from my own observing experience and preference.

Use the attached star chart and the following Guide as an excellent reference for your next star party itinerary, or a beginning for further study into the thousands of objects visible in this part of the sky. Truly these extensive *Constellation* Study Guides will most definitely put your computerized telescope to work for you in the most efficient and enjoyable way possible! As a matter of fact, MANY telescope users are now programming their own "Tours" based on these guides, using each constellation as a separate GO TO Tour for the GO TO star program library that can be added in or deleted through the main edit screen on your PC or MAC computer.

We hope you enjoy these comprehensive GUIDES to touring the constellations via your AutoStar and other computer-driven telescope. Each new installment is complete with diagrams, charts and illustrations that you will find nowhere else. Please let us hear YOUR feedback and your observations of each and every constellation after YOU have toured its vast reaches of our skies!

YOUR AURIGA CONCISE DIRECTORY OF INTERESTING OBJECTS –

In addition to our regular listing of a few selected objects, I have included the complete abstract listing for the ten (10) nice galactic star clusters that are viewable in this constellation via modest instruments; that listing will be found at the end of this concise description of our TOUR.

For our visit into the borders of the celestial charioteer I have chosen the finest (or most interesting) 11 objects in this AURIGA "GO TO" tour (as with all guides, all objects listed below will be visible in most telescopes (some naked eye) from the smallest to largest telescopes; of course larger apertures may "show" an object a bit closer and "better," but frequently a wide field and low power view is more desirable than aperture for FINDING the objects initially. Indeed, I strongly encourage you first FIND the target object, or its approximate location through your GO TO function with your lowest power and then - once IDENTIFIED positively - move up slowly in steps with magnification if necessary. Remember, not all objects "like" magnification. Sometimes better "field of view" (such as the wonderful wide fields provided by smaller telescopes) is desired over light gathering (like an 8-inch) and magnification.

The rule for determining "optimum magnification" is that: 1) too low power results in sky background glow detracting or diminishing the contrast against the deep sky object; 2) too high magnification darkens BOTH the sky background AND the object; 3) medium magnification can be achieved at which you have MAXIMUM contrast between the object and its darkened background sky. I have found through three decades of direct observing that about 15x per inch aperture. for deep sky observing is PERFECT for most objects. That being said, always remember that DOUBLE or multiple stars require whatever power you can crank out....the seeing conditions are the limiting factor here. For a complete discussion on magnification and how it applies to YOUR telescope, visit my under the GUIDES section of this website._

For my complete and comprehensive discussion regarding seeing conditions and sky transparency, see also this topic under the GUIDE to seeing and transparency from Arkansas Sky Observatories.

With all deep sky objects, avoid attempting to observe when the moon is in the sky, even a very thin crescent, as its brightness in the sky will overshadow the very dim contrast afforded by even the brightest deep sky object; if you see the object at all against moonlight, you will NOT see the subtle outlying areas or the full detail of what is presented.

Also, as I always suggest with all of the "GO TO" tour constellation lists, a good star atlas and/or chart which will list all the finest objects, constellation-by-constellation. One very handy reference guide is the *PETERSON FIELD GUIDE TO THE STARS AND PLANETS*, which features complete lists with declinations, right ascensions, magnitudes, and all pertinent information for you to expand your observing horizons beyond this brief GUIDE. For multiple stars and many listings of the finest deep sky objects, the classic work, *Burnham's Celestial Handbook*, Vol. One is highly recommended.

Use the attached star chart (above) and the following Guide as an excellent reference for your next star party itinerary, or a beginning for further study into the thousands of objects visible in this part of the sky. The chart gives the outline of the major constellation as well as the approximate locations of all objects discussed in this guide.

Following is the concise object list for your "GO TO" tour of AURIGA; you may wish to find the majority of the objects from the Sky Library (for example, you can easily go to the beautiful and rich star cluster Messier 37 if you pull up "Object/Deep Sky/Messier/..then type in '37'...." and then press "Enter", followed by "GO TO" to access this very delicately star-laced gem. On the other hand, if you want to experiment and become a "better computer user" try entering the exact R.A. and DEC coordinates (given in the listing below) of that same object as described in the HELP section of your computer Sky program.

You will access your first GOTO target - (usually the brightest star in each constellation) - via the command "SETUP / OBJECT / STAR / NAMED....and scroll to "Capella"", then press "Enter" and subsequently "GO TO" to move your this bright star. Remember also that many distinctive objects are sometimes listed among the "named" objects. So, likewise for that object you might merely go to SETUP/OBJECT/DEEP SKY/NAMED....and then scroll alphabetically to the "common" name of the object if you are not already there; press "enter" and then GO TO and your scope is off and running! For Auriga, there are NO objects other than the brighter stars that are listed as common "named" objects in the Autostar library by name.

You may also access the constellation by: SETUP/OBJECT/CONSTELLATION/"Auriga"..... Enter....GO TO, which will take you close to the central position of the constellation's boundaries.

OBJECT 1: bright star - CAPELLA (alpha Aurigae) - R.A. 05h 13" / DEC +45 57 - Magnitude: 0.1 - deep yellow

OBJECT 2: triple star! - Theta Aurigae - R.A. 05h 56' / DEC + 37 13 - Mags: 2.7, 7.5 & 11, nice target!

OBJECT 3: double star - 5 Aurigae - R.A. 04h 57' / DEC + 39 19 - Mags: 6.0 & 9.7 @ 3.2" apart - nice stars!

OBJECT 4: great easy double - 41 Aur - R.A. 06 08' / DEC + 48 43 - Mags: 6.8 & 6.1 - good in smaller apertures!

OBJECT 5: tough test double - 54 Aur - R.A. 06h 36' / DEC + 28 19 - Mags: 6.1 & 7.7 - test for 5", 8" – tough

OBJECT 6: " test double - Burnham 1053 - R.A. 05h 50' / DEC + 37 20 - Mags: 7.5 & 9.3 - at limit for 3" !

OBJECT 7: variable star - RW Aur - R.A. 05h 05' / DEC + 30 20 - irregular, Mag 9-12, very erratic!

OBJECT 8: variable star - SS Aur - R.A. 06h 10' / DEC 47 46 - Dwarf nova - Mag 10 to 15, very peculiar star

OBJECT 9: galactic cluster - Messier 36 (ngc1960)
- R.A. 05h 32' / DEC + 34 07 - Mag: 6.3, 60 stars

OBJECT 10: galactic cluster - Messier 37 (ngc2099) - R.A. 05h 49' / DEC + 32 33 - Mag: 6.2, 150 stars

OBJECT 11: galactic cluster - Messier 38 (ngc1912) - R.A. 05h 25' / DEC + 35 48 - Mag. 7.4, 100 stars (fine!)

Plus:GALACTIC CLUSTERS IN AURIGA:

Following is an abbreviated listing of the ten (10) brightest and most interesting open, or galactic, clusters in the constellation of Auriga; this area of the sky if VERY rich in wonderful star fields for scanning with the very low power and wide field instruments as well as a standard pair of 7 x 50 or 10 x 50 binoculars, since it is located just on the edge of the dark and rich winter Milky Way skies. The listing following gives the NGC #, the R.A. and DEC of the object, the Magnitude, Size, # of stars and a brief and very concise description and/or notes as necessary:

ngc	RA	DEC (+)	MAG	SIZE ('arc)	#stars	
1664	04 47	43 37	7.5	15	40	not much
"cluster" effect; good in very wide field views						
1857	05 17	39 18	8.6	9	45	very nice and
compact; use med. mag.						
1893	05 22	33 21	8.0	12	20	only a few
stars, pretty scattered, low power						
1907	05 25	35 17	9.9	5	40	very faint and
small, compact....med.-high power						

1912 05 25 35 48   7.4   20   100 M38, very large and scattered, best in wide field
1960   05 32 34 07   6.3   12   60 M36, brighter stars over wide area; low powers
2099   05 49 32 33   6.2   20   150 M37, the best of the bunch, many tiny stars!
2126 05 58 49 55   9.8   7   30 very small and compact; medium power
2192 06 11 39 50   10.9   6   30 similar to above, small, faint stars, nice object
2281 06 46 41 07   6.7   17   30 very large with few stars; brighter members

A VISUAL GUIDE TO OUR DEEP SKY OBJECTS IN AURIGA

Object 1 - Our "Starting" Bright Star - "CAPELLA" (alpha Aurigae)

Known as "The Goat Star" (because our charioteer is carrying a bunch of goats, remember?), this star also has signified the head of Auriga on many occasions, particularly if you subscribe to the sea-god chariot being drawn by the curiously-large sea-horse myth. This beautiful golden star is among my favorites, perhaps because - like Arcturus is a the signal of impending springtime - Capella is the harbinger of the crisp, cool skies of Autumn in the northern hemisphere. As mentioned earlier, Capella reaches its highest point in the sky (culmination) at midnight in the second week in December. At magnitude minus 0.06, it is the 6th brightest star of the sky, only 45 light years away from our own sun. Capella is actually a VERY complex multiple star system. The primary star "A" is a binary that cannot be seen with telescopes, only spectroscopically; this pair has another component "H" which is a red dwarf 10th magnitude star 12' due SE from Capella; if you look VERY closely at this faint star (use the

5" and larger telescope), this star will also split into two stars, in a SE to NW direction (Position Angle 137 degrees), nearly 3" arc apart. Have fun with this one!

Object 2 - Triple Star - Theta Aurigae
The primary "A" star of this triple system is a bright 2.7 magnitude, so it is easy to locate in the finderscope if you are not using the GO TO function of your computerized telescope. The system is more than twice the distance as bright Capella and is known as a "Silicon star", with a great deal of its chemical "fuel" now fused into a silicon state.

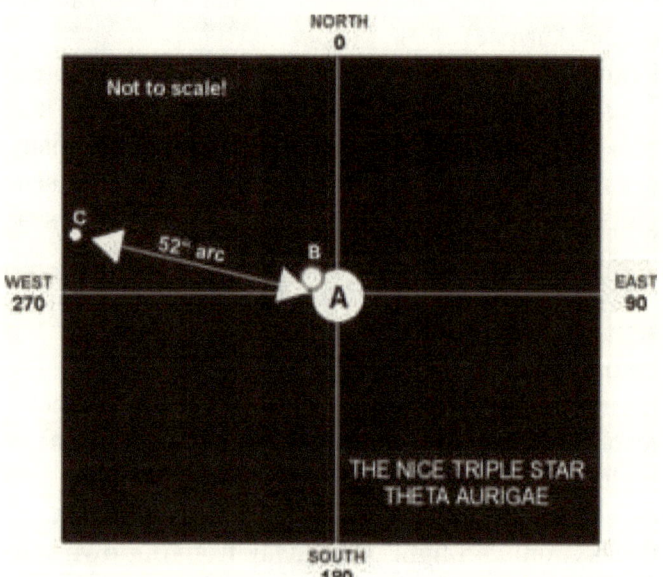

Look for the "B" companion, magnitude 7.4, in Position Angle 318 degrees, or about NW from the primary "A" star. It is just over 3" arc away, so it will required medium to medium-high power to see this star in contrast to the very bright primary. The third star "C" is in Position Angle 300 degrees (almost in line with A and B) but is a much farther 52" arc (about the size of Jupiter in the same eyepiece), and is considerably fainter at magnitude 11. Thus, although the 3" scope can theoretically show all three of these stars, the 5" and larger are recommended for best success.

Object 3 - 5 Aurigae - A very nice and easy double star for all telescopes!
Currently this double star is separated by nearly 3.4" arc, with the stars in a nearly East-West orientation....no excuses on this one folks. It is a good and fun star. The primary is magnitude 6.1 and very bright white, while the fainter secondary star is magnitude 9.5 and more yellow-orange.

Object 4 - Another Good Double - 41 Aurigae - Here is one for everyone!
This is my "so-you're-frustrated-and-need-an-easy-object" double star. This one will take the stress off after trying to glimpse the multi-faceted nature of Capella. 41 Aurigae is relatively bright and easy double for all telescope, provided that you are assured you are looking at the right star (there are several stars of similar magnitude nearby, but none with a companion!). This yellowish star of magnitude 5 has a 7.7 magnitude companion nearly 8" arc away (easily resolvable) near DUE NORTH....again, no excuses on this one! Just observe it for a long time, take a deep breath, and then go on to more challenging objects! You've earned the break.

Object 5 - Test Double for 5" to 8" Telescopes –
54 Aurigae

Here is an excellent test object for the 5" telescopes under very steady conditions. We have not discussed the "resolution limit" of telescopes much here, the general standard is termed "*Dawes' Limit*" or "*Dawes' Criteria*" based to entirely empirical data compiled from many telescopes observing a wealth of stars exhibiting various degrees of separation. Suffice it to say, the if your divide your telescope's aperture (in inches) into the value "4.54", this is as CLOSE a double star in arc seconds (") as you can resolve with your scope. Thus, for telescopes of various apertures:

2.5"	inches	=	1.81"	arc
3.0"		=		1.51
3.5"		=		1.30
4.0"		=		1.13
5.0"		=		0.91
6.0"		=		0.76
8.0"		=		0.57
10.0"		=		0.45
12.0"		=		0.38

In my over-thirty years experience in direct astronomical observation, I will attest to *Dawes'* accuracy in regard to these actual "reachable" numbers. They are very realistic and precise for all apertures. There are some variation with very good and very bad optics as well as with seeing conditions and/or the type of telescope used; Maksutovs general will resolve closer stars inch-for-inch than a Schmidt Cassegrain, while a "pure-aperture" refractor will surpass resolution in most cases of both of those instruments.

The star 54 Aurigae is at present 0.91" arc, just at the Dawe's limit for the 5" scope; however, my latest observations on a night of "fair-to-moderately poor" steadiness failed to reveal the duplicity of this star with powers ranging from 127x through 404x. It does split readily in the 6" Unitron refractor, and very easily with the 10" LX 200. It is a test object, however, for the 8" on the same night....

However, I do believe that a good quality 5", well collimated Maksutov will split this star; the primary star "A" is at magnitude 6.0, so it is easy to isolate; increase the magnification to at least 200x and even more on a very steady night. The companion 8th magnitude star, "B", will be found NE of the brighter star and VERY close, nearly "touching" the image (the "Airy disk") of the brighter star. In a well collimated telescope keep in mind that many times a fainter star will appear embedded in the faint rings known as "Airy diffraction rings," so if you do not first see this faint and close companion, be sure to examine the closer ring(s) closely...it might be hiding in there!

Object 6 - Now....here is one for the 3" scopes!
Double Star Burnham 1053
Once again, you will access this star either by using the R.A. and DEC manually entered via the hand controller, or by the "old fashioned" setting circles on your scope (they actually work very well....and it's fun!). Burnham 1053 is a very close star for the 3" to 5" size range, separated by 1.4" arc, right at the Dawe's limit for the 3" scope. The primary "A" star is magnitude 7.5, so it is not the brightest of our Auriga objects by a long shot. Center that yellowish

star and increase the magnification to about 200x and look for a faint 9.6 magnitude star near DUE NORTH (a bit to the west) and very close. You will require a very steady night and very high magnification (take it on up if the air will hold it!) in the smaller of the two scopes.

Object 7 - A Very Unusual Variable for Amateur Telescopes - **RW Auriga**
Here is an excellent and fairly bright variable star that is one of the most unusual known. RW Auriga varies between magnitude 9.0 and 11.6 with no particular period; it is one of many stars known as "T Tauri" variables which are thought to have NOT reached equilibrium yet and are still settling down from their original formations!

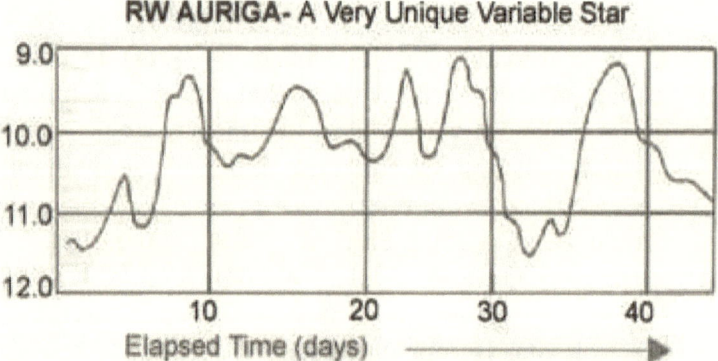

Looking at the light curve, you can clearly see that there are distinctive "peaks" at which RW Aur surges upward in brightness, making it an easy object for the 5" and larger telescopes; as a matter of fact this star can be studies throughout its erratic light variations with such telescopes. Although the star can vary dramatically in less than one day, note that there is some indication of a "periodicity" or frequency of about 10 days at which the star may or may not surge to maximum. Observations should be made at EVERY opportunity as this is one of those stars that can change brightness literally within hours!

As with all variable star observing, the ideal source for information, comparison star charts, and to report YOUR valuable observations (badly needed on these irregular stars, by the way!), is the American Association of Variable Star Observers (AAVSO - www.aavso.org) in Cambridge, Mass. The charts for nearly every variable star - as well as "new stars" that appear from time to time - can be downloaded off the internet into a file in your computer; save the file and bring it up.....the image will be huge. Resize this image to fit your page, resave, and then print for a good chart to use at the telescope!

To make estimates on this peculiar star, use the AAVSO "t" chart found at: https://www.aavso.org/apps/vsp/ . Note for these charts, simply type in the NAME of the variable at top to generate your choice of chart.

This is the narrow-field chart which shows faint comparison stars, and a "locator" chart from the AAVSO is not available for this star. Also, note that this is the AAVSO "standard" chart and not the "reversed chart"; the reversed editions of the charts are newer and are matched to the reversed field of view of the Maksutov and Schmidt Cassegrain telescopes, which result from the diagonal prism or mirror.

Object 8 - A Dwarf Nova Variable - SS Auriga - Faint, but an excellent star for the 5" and larger scopes!

The star SS Auriga is an incredible object for amateurs with moderate to large aperture telescopes. It varies in light from brightest at about magnitude 10.5 to the very dim 15th magnitude minimum. To locate the star as well as making estimates when it is at its brightest, use the following AAVSO "a"chart:

https://www.aavso.org/apps/vsp/ . Note for these charts, simply type in the NAME of the variable at top to generate your choice of chart.

however, when faintest, the "g" chart which will give stars to compare down to 14th magnitude.

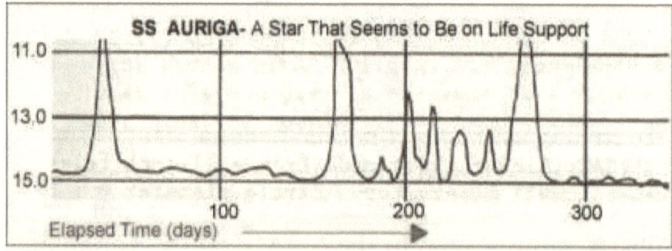

For observers with 5" scopes and large, this is an exciting star to monitor on a regular basis....SS Aurigae's brightness will fluctuate unexpectedly from brightest to dimmest, hence it is very important that observers constantly monitor this star. Your observations are quite valued by the AAVSO to "fill in the gaps" as to the true nature of this curious star and others like it. Note that from the light curve above, the star will usually have a spiked outburst about every 55 days. This is very similar to the star SS Cygni, also a dwarf novae and which was discussed in the ASO *CYGNUS* Constellation Guide .

Object 9 - Beautiful Galactic Cluster Messier 36

The next three objects of your AURIGA "GO TO" TOUR are all beautiful galactic star clusters found in the southwest "corner" of this constellation. The photograph below, taken through the wide field astrograph at Lowell Observatory, demonstrates the proximity of these beautiful clusters to one-another. It ALSO points out clearly that aperture is NOT needed for your best views of these clusters (although my favorite memory of Messier 37 with its very faint stars was, indeed, through a 12" Meade LX 200). As a matter of fact, this photography was taken through Lowell's sky patrol camera which is fitted with a 5" lens! Note the richness of the star field in this immediate area, as the Orion arm of the Milky Way stretches through southwest Auriga, filling it with tiny images of delicate stars. South is at the TOP of this photograph.

Messier 36 is probably the most compact-looking of all the three, and is, indeed, the smallest in terms of observed area, at only 12' arc across....half the size of the other two. It contains about 60 stars which range from magnitude 8.7 to less than 13th; the brightest of these can be seen in scopes smaller than 3"; this is an ideal object for a 3 inch scope. Use very low power and a wide field of view to appreciate the beautiful star fields surrounding all three of these wonderful clusters!

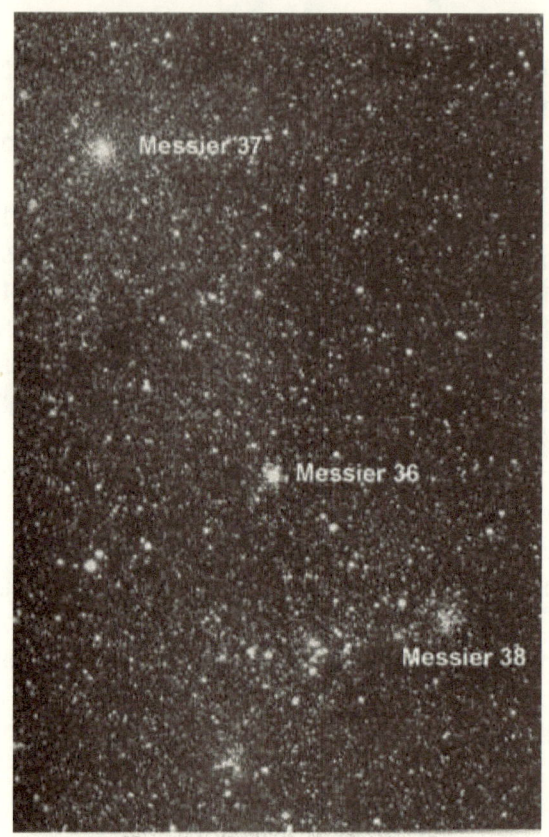

Messier Objects in Auriga:
Beautiful Galactic Clusters M-36, 37, & 38

<u>Object 10</u> - Number Two of the Trio - Galactic Cluster Messier 37
Other than the "Wild Duck" cluster (Messier 11), this is my favorite of all galactic star clusters. In a pair of good binoculars on a dark and moonless night, this cluster appears as a large glow of light, unresolvable in any optics less than about 2" diameter. However, even the smallest of telescopes at very low power can begin to see many of the

nearly 200 stars embedded in this truly spectacular cluster, the northernmost of the three in Auriga. In 3" and larger telescopes with very wide fields, one is immediately impressed with what appears to be hundreds of tiny diamonds scatter against a blanket of black. Just look at the photograph above and note the dense concentration of very faint Milky Way stars that pass ever so close to M-37! The stars are nearly uniformly between magnitude 10 and 12, but some brighter members do exist. I think that it is this uniform brightness of well over 150 stars that gives it such a dramatic view. Look carefully near the very center of this cluster....right there is the 9th magnitude **RUBY STAR** of Auriga, a red giant that stands out among all the other yellowish members of this cluster!

Objects 11 - The Final Triplet of Galactic Clusters: Messier 38

Messier 38, in my opinion is "better than M36.....but not as spectacular as M37." As the northernmost of the three big galactic Auriga clusters, M38 is positioned slightly out of the thickest and most star-dense part of the Milky Way stretching through this constellation. However, if you use VERY wide fields and low magnifications and scan just to the east and north of this star a couple of degrees, you will not be disappointed at all, as the "clumps" of stars - some of them star clusters themselves - seemingly never end. M-38 is the same size as Messier 37 (20' arc, almost as large as the moon appears) and contains about 100 stars....again, more than M-36 but not as many as M-37! In low power, look for the distinct shape of a CROSS in this cluster of stars, with each arm having a double star at the end! A fantastic object for public attention at your next star party! There is one very yellow (giant) star of magnitude 7.9 in the cluster which definitely stands out. If you are using the 5" and larger telescopes, look just to the southeast of M-38 for the tiny cluster ngc1907, a beautiful "jewel box" contain some 40 very faint stars!

WANDERING ABOUT....YOUR NEW "USER OBJECT" IN AURIGA

In the constellation guide "GO TO" ARIES (discussed with *ARIES*), we discussed three stars that were known as the "*Runaway Stars*" from their point of "birth" in the Orion Nebula. These three stars - 53 Arietis, MU Columbae and now AE Aurigae - were somehow and for some strange reason "ejected" after forming in the central region of the famous nebula some 2.7 million years ago and all three are flying away from one-another in totally different directions! (see the diagram following)

Of the three Mu Col is the brightest (5.2), followed by AE Aurigae at 6th magnitude (although it is an irregular and erratic variable, from 5.4 to 6.2).

AE Aurigae is located at coordinates: R.A. 05 13 and DEC + 34 15, and can be found through the following diagram...NOTE: if you use stars # 14, 16, 17 & 19 as a "line", AE will make a northwestward right angle from #19 Aurigae!)

146

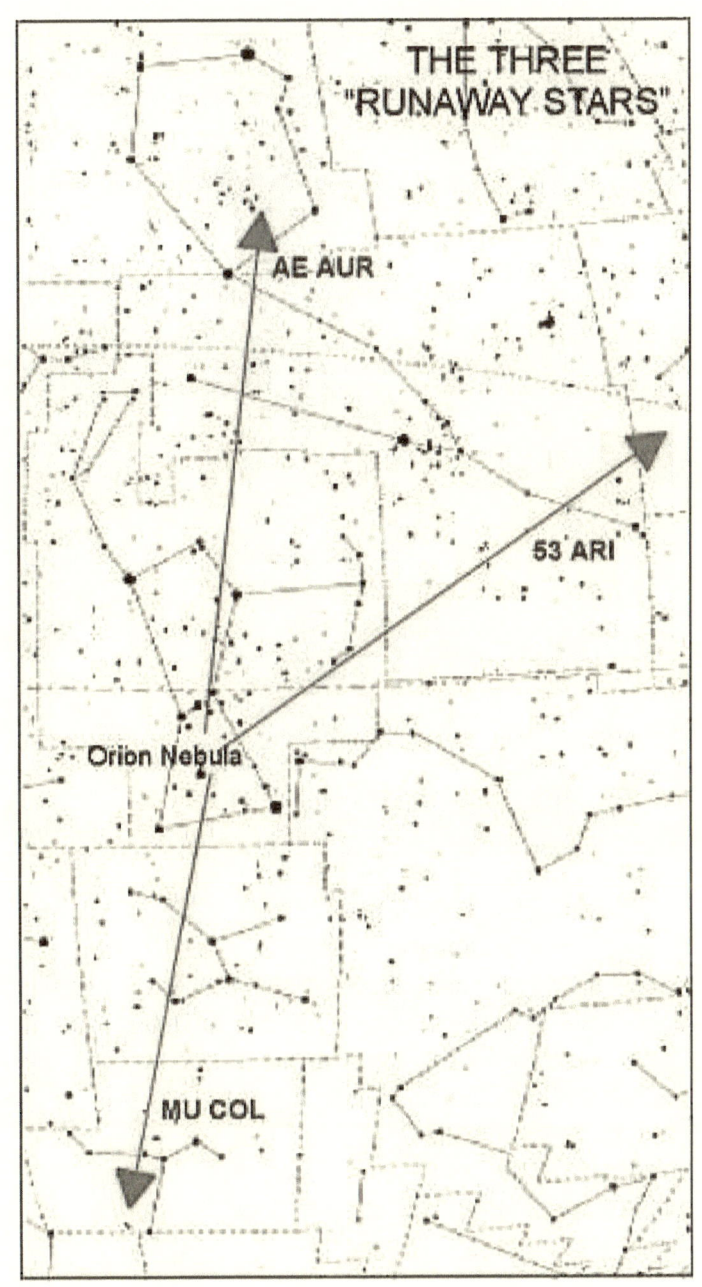

THE THREE
"RUNAWAY STARS"

AE AUR

53 ARI

Orion Nebula

MU COL

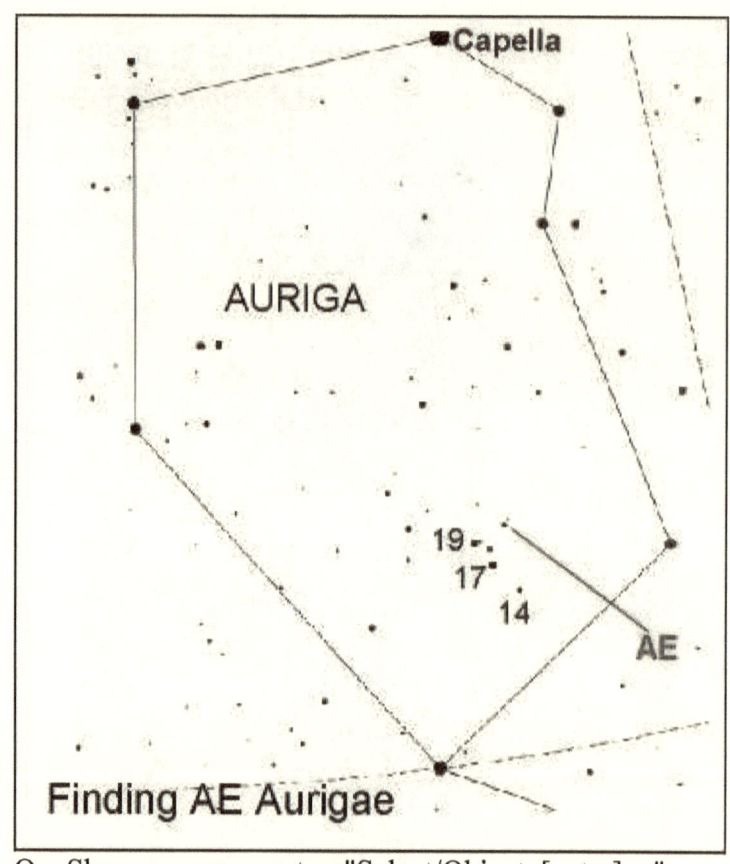

Finding AE Aurigae

On Sky program go to: "Select/Object [enter]...." scroll down to "User Object" [enter]. Now enter the coordinates given above for "AE Aur", using the number keys (on AutoStar). After entering the coordinates and pressing "Enter" yet again, scroll down one and you can list the magnitude of the object as "6"[Enter].

With the addition of this star, you now have a runaway star in the collection of celestial curiosities for your next star party or family outing with the telescope!

* * *

Be sure to read about nearby PERSEUS (Constellations, Vol. II), the fabulous Greek hero who rescued poor Andromeda from her father's tethers to the sea monster Cetus, and who seems to have more than a casual liking to both her and her mother, Cassiopeia. As always, you can learn more about this philandering Fabio in our "GO TO" tours constellation guide.

Good Observing and may the stars serve as your sentries as you explore the frontiers of space!

Hevelius' *Auriga*, 1687

Even a small computerized telescope can become a
powerful tool for viewing the thousands of celestial
objects within our 88 constellations.
(*Doc Clay's Supercharged ETX 125*)

Chapter Six

BOOTES
With discussions of objects in
Corona Borealis, Coma Berenices and Serpens

Entering the Domain of "THE HERDSMAN"

In this constellation guide installment, "GO TO BOOTES"- of our constellation study guides for all GO TO telescope users - we will explore the absolutely remarkable array of fine globular clusters, galaxies and doubles stars found in this group of constellations. As with all of our GO TO guides", it features a "start" with an easy GO TO to the bright star **ARCTURUS**, and then proceeds through the many fine examples of double and multiple stars, galaxies, clusters and interesting objects within reach of YOUR telescope! From bright starry globular clusters to beautifully-colored and complex multiple star systems, discussions along the way tell you what to expect from each telescope size and type. All objects will be discussed with exact descriptions of what the viewers using various aperture telescopes should expect to see...and what to NOT expect to see!

The finest stellar and deep sky objects in these constellations will be featured....and - yes - there will be something for everyone and every telescope...Even naked eye and binoculars when appropriate!

Discussed are useful magnifications for EACH GO TO object, what type of night and conditions are needed to see certain details, double stars that can be resolved in each telescope model, and much, much more. It is your complete GUIDE for your deep sky observing pleasure and a very handy tool for use at your next star party!

Needless to say, it WILL put your computerized telescope to work for you in a most efficient and enjoyable way!

I hope you will enjoy these comprehensive guides to "Touring the Constellations" which feature most of the interesting constellations, complete with diagrams, charts and illustrations . Please let us hear from you with summations of YOUR observations through these constellation tours!

Introduction to Bootes

The constellation Bootes (pronounced "bo-o-tees" is marked by the remarkably beautiful and bright orange yellow star **ARCTURUS**, the *"heralder of spring."* But a discussion of the absolutely fascinating array of deep sky objects in Bootes would not be complete without including the many similar objects that rest just outside of its boundaries....in the smaller constellations of **Corona Borealis**, **Coma Berenices** and **Serpens**.

A finder chart for locating many of the GO TO objects in the constellation of Bootes follows; if using a computer planetarium program, you are encouraged to plot the objects on your screen for higher resolution than this chart provides.

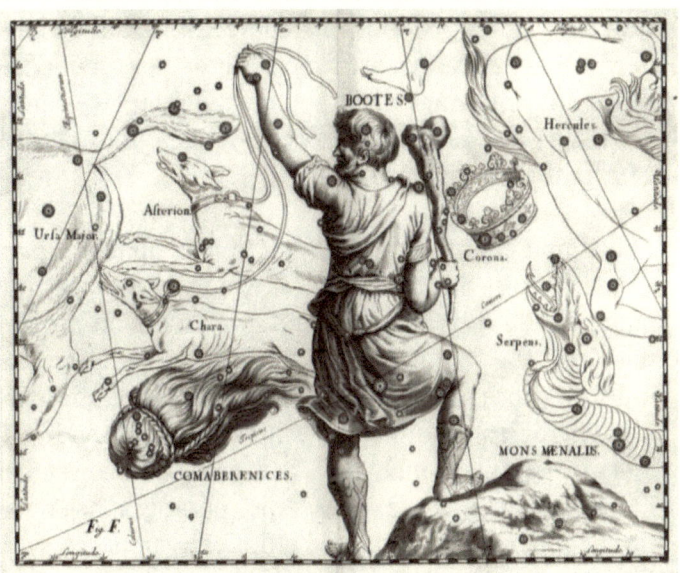

Beautiful artwork of Bootes from
Johannes Hevelius Atlas

It really takes some imagination with this group of stars to see the character imagined by the ancient Arabian skywatcher as a *"bear herder"* and in later cultures as merely a *"herdsman of animals."*

Perhaps not so difficult to at least "imagine" is the *"Berenice's Hair,"* or Coma Berenices, the bright sprinkling of stars throughout this "mel 111" cluster of stars resembling the faint golden locks of hair immortalized in the heavens from *Berenices II* of Egypt the queen of *Ptolemy III* (246 B.C.).

As Ptolemy had gone to a treacherous battle far away, his safe return was begged by Berenice in exchange for her legendary hair locks which had been placed as sacrifice in the temple of *Aphrodite*.

153

Only after the hair had vanished, did the ancient star watchers of Egypt look up and realize that it had been magically transformed into the delicate spattering of stars in the constellation we now recognize as "Coma Berenices", a star group about 250 light years distant.

There are about **35 stars** that make up the main Coma group, the brightest of which are about 5th magnitude. Only in wide field binoculars on a dark night can the true beauty of this cluster be seen. but it is not the star groups that the constellation is known for astronomically.....it is the thousands of distant galaxies that are seen past those stars, literally filling large telescopic photographs with field after field of spiral and elliptical wonders.

On the opposite side (east) of Bootes is another small star pattern, the "*northern crown*" or **Corona Borealis**, a very pretty ringlet of stars marked clearly by the crown jewel, *COR CAROLI.*

YOUR BOOTES
COMA BERENICES
CORONA BOREALIS CONCISE DIRECTORY

We will concentrate on 15 objects in THREE-CONSTELLATIONS (four if you count one object in "Serpens")tour; all are in reach of telescopes 3- to 8-inches aperture, yet each telescope will demonstrate uniquely different and challenging aspects of the objects. In addition to the 15 finest objects, there are literally hundreds of faint galaxies that are visible in most scopes as well as hundreds of double and multiple stars, stars of curious colors and motion. It is in this general area, and in the constellation of *OPHIUCHUS* (see Constellations/ Ophiuchus) that more galaxies can be viewed from Earth than at any other point in its orbit.

As with all of our GO TO tours, I recommend good a good star atlas and/or chart/and PC sky program which lists the finest objects constellation-by-constellation; if you cannot access any of these objects (or those that are not listed in this tour), you can access directly from coordinates - Right Ascension (RA) and Declination (DEC) of any known object via the PC program or telescope keypad. This process varies from program-to-program, but as an example, you merely need to hold down your MODE (back) key on the AutoStar for three (3) seconds and the RA and DEC coordinates appear for the telescope. Merely press "GO TO" and the cursor appears prompting you to enter the Right Ascension of the object if it is NOT listed among the objects in the AutoStar library; once the RA is entered, press "Enter" and the cursor once again prompts for the Declination coordinates (these coordinates for epoch 2000) are found in all good observing guides). Once those are entered, merely press "GO TO" once again and your computerized telescope will slew to the position of the object!

The following constellation guide to objects at the end of this tour will describe all the details of each object and provide specifics as to visibility of that object in YOUR telescope model.

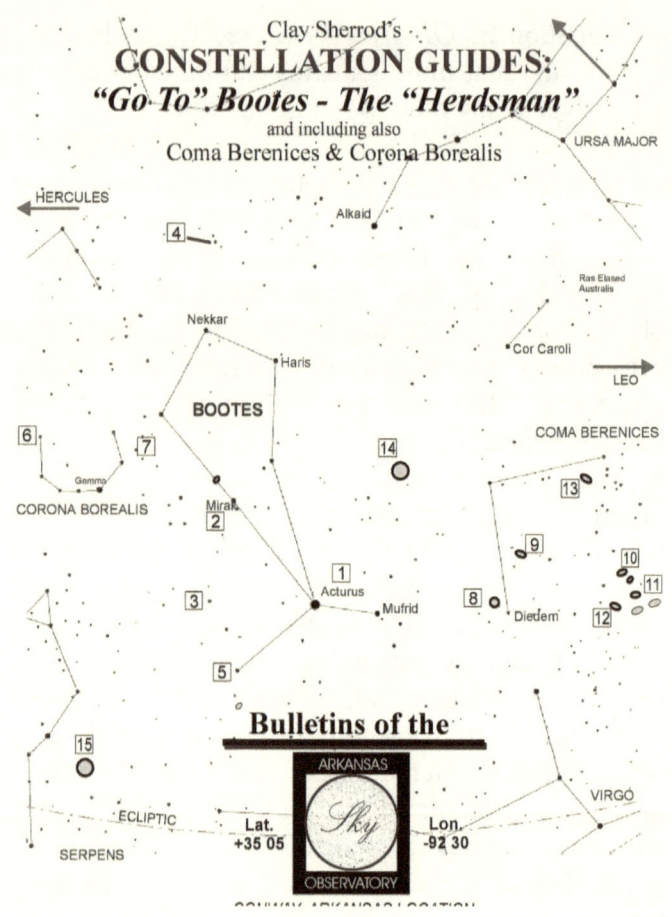

Clay Sherrod's
CONSTELLATION GUIDES:
"Go To" Bootes - The "Herdsman"
and including also
Coma Berenices & Corona Borealis

URSA MAJOR

HERCULES

Alkaid

[4]

Ras Elased
Australis

Nekkar

Haris

Cor Caroli

LEO

BOOTES

COMA BERENICES

[6]   [7]   [14]
                              ○

Gemma   [13]

CORONA BOREALIS   Mirak   [9]   [10]
                     [2]   [11]

                     [1]   [8]○   [12]
                     Acturus   Diedem
[3]   Mufrid

[5]

Bulletins of the

[15]
○

ARKANSAS

VIRGO

ECLIPTIC   Lat.   *Sky*   Lon.
             +35 05   -92 30
SERPENS
             OBSERVATORY

The star chart seen above will get you started, as it demonstrates the relative positions of all objects in this "tour" to the conspicuous stars outlining the distinct figures of our three constellations of this TOUR.

Following is the complete 15-object list for your "GO TO TOUR" of **Bootes**, **Coma Berenices**, **Corona Borealis** and one object in **Serpens**; you may wish to find the majority of the objects from the AutoStar or planetarium program library (for

156

example, you can merely pull up Messiers 5 or 3 by going to "Object/Deep Sky/Messier Object/M-5....enter....GO TO" or...if you want to experiment and be a "better computer user", try entering the following coordinates (provided in the list directly following) as described under MODE above.

OBJECT 1: bright star - ARCTURUS (alpha Bootis) - R.A. 14h 13" / DEC + 19 27- Magnitude: -0.1

OBJECT 2: beautiful double star - MIRAK (epsilon Bootis) - R.A. 14h 43' / DEC + 27 17 - Magnitudes: 2.4 & 5.0

OBJECT 3: beautiful double star - XI Bootis - R.A. 14h 49' / DEC + 18 18 - Magnitudes: 4.8 & 6.8

OBJECT 4: tough, 8-inch test double! - 44 Bootis - R.A. 15h 02' / DEC + 47 51 - Magnitudes: 5.9 & 7 ? variable

OBJECT 5: a great double for all! - Zeta Bootis - R.A. 14h 49' / + 19 18 - Magnitudes: 4.8 & 6.9

OBJECT 6: double star test for 4 inch! - Sigma Cor. Bor - R.A. 16h 13' / DEC + 33 59 - Magnitudes: 5.7 & 6.7

OBJECT 7: double star test for good 6-inch Eta Cor Bor. - R.A. 15h 21' / DEC + 30 28 - Magnitudes: 5.7 & 6.0

OBJECTS 8: (Coma Berenices) 2 globular clusters - Messier 53 (ngc5024) and ngc5053 - R.A. 13h 11' / DEC + 18 26 (super!!)

OBJECT 9: (Coma Berenices): "black eye" galaxy - Messier 64 (ngc4826) - R.A. 12h 54' / DEC + 21 57 (a really cool object!)

OBJECT 10: (Coma Berenices): galaxy - Messier 85 (ngc4382) - R.A. 12h 23' / DEC + 18 28

OBJECTS 11: (Coma Berenices): 3 galaxies - Messiers 98 (ngc4192), 99 (ngc4254), 100 (ngc4321) - R.A. 12h 11' / DEC + 15 11
OBJECT 12: (Coma Berenices): galaxy - Messier 88 (ngc4501) - R.A. 12h 30' / DEC + 14 42
OBJECT 13: (Coma Berenices): the famous "needle" edge-on galaxy - ngc4565 - R.A. 12h 34' / DEC + 26 16 (a must-see!!)
OBJECT 14: (Canes Venatici): globular - Messier 3 (ngc5272) - R.A. 13h 40' / +28 38 (this is the one you don't want to miss!)
OBJECT 15: (Serpens): wonderful globular - Messier 5 (ngc5904) - R.A. 15h 16' / DEC + 02 16 - *THE FINEST GLOBULAR*!

A VISUAL GUIDE TO OUR OBJECTS IN THIS CONSTELLATION SELECTION –

Object 1 - Bright Star Arcturus (Alpha Bootis)
Unknown to most, this is the brightest star NORTH of the celestial equator, shining a beautiful deep yellowish red color, even when low on the horizon. To the ancient people of northern hemispheres, the first rising of Arcturus seen at dusk, just when enough darkness had fallen, signified the "heralder of springtime." Even more fame for this star (Arabic for "*the Bear's Guardian*") came recently during the opening of the **World's Fair in 1933** to signal the new era of technology (boy, did they predict that one right!). To open this exposition in Chicago, Arcturus' light was telescopically focused to a photoelectric cell, thus throwing on the power via its switch to the entire gala affair! At 40 light years distant, the very light that opened the fair had left Arcturus in 1893 in the year of the last previous such fair held in that city!

At magnitude -0.1, Arcturus is one of the few stars that can be found with your GO TO computer during daytime! After your eyes get accustomed to the bright background light of the finderscope, it is clearly visible when nearly overhead and far away from the sun's glare. Arcturus is huge - about 20 million miles across - and well over 100 times more luminous than our sun.

Object 2 - Bright Double Star MIRAK (epsilon Bootis)
This double star - at a separation of 2.9" arc and magnitudes of 2.4 and 5.0 - would be thought to be relatively "easy" double star to split with nearly all of our telescopes. However, this one is a tough, tough test for the 4-inch scopes and not at all resolvable with smaller scopes. With the 6-inch, the star is more easily resolved at very high power on a good night, but it is still not a simple task. With this scope and the 8-inch, look for a wonderful color contrast: yellow-orange for the brighter star and blue for the fainter one. Much magnification is required to see both the color and the clear separation of these two stars.

Object 3 - Double Star XI Bootis
This star is one of the finest doubles for all telescopes. At 7" arc separation, you can clearly see both components of this very reddish pair of stars. With a combined magnitude of about 4th, they are an easy target and a nice "show" object. With an orbit of nearly 150 years, this is about the widest you will ever see these two stars separated. Look for the fainter star just a bit NNW of the brighter "parent" star. Medium power should be used with small scopes, about 80x to 100x in all other scopes.

Object 4 - An acid test for your 6- or 8-inch telescope - Double Star **44 Bootis** So you think you're scope is perfectly collimated....great optics....good eyesight....running with the big dogs. So aim that larger light bucket over to 44 Bootis and let's see what you can do! The main star of this double is almost exactly like our sun....same size, luminosity, color. However, at a distance of ONLY 750,000 miles is another star orbiting that sun. This is the time to attempt to split this one, folks, having just now reached it APHELION to the east of the primary star, the little secondary star is a full 3" arc DUE NE of the main star. But don't let that distance fool you...it is a tough one. The second star is a bit variable and erupts from time to time. Note that the apparent separation is again closing between these two stars; in 2040, they will once again only be about 0.3" arc apart.....so see it while you can!

Object 5 - Great double star Zeta Bootis - A super double star test for the 4-inch
This is a gorgeous double star, with components of magnitude 4.4 and 4.8 and a separation of about 6" arc. Because of their similarities in magnitude, this is a relatively tough star to split with the 3-inch, but is easy with the 4-6 inch scopes and high power. On a very steady night, however, the 4-inch should open these up. Although many books have reported that it is only resolvable with a 4" telescope or larger, I have clearly split this star with a Questar 3.5" but have unfortunately not had an opportunity to try with a smaller telescope....results and reports on this, please!
Object 6 - Double Star in Corona Borealis - Sigma Cor Bor

At a relatively bright magnitude 5.4 and the wide separation of about 8" arc, this is a very pretty and "easy" double star for all of our telescopes. Medium powers (about 15x per inch) are recommended for the entire range of aperture here. Both are almost equal colors. An added bonus for those with the 8-inch....look about only 12' arc to the SOUTHWEST from the main star....here you will see a very red star **LTT 14836**, a very classic *red dwarf* star that, even with its faintness, is unmistakably red in color.

Object 7 - Double Star for small APO Test - Beautiful object for all!
This star, ALULA AUSTRALIS, is actually in the constellation of Ursa Major (one of its back paws) and was featured in the previous constellation TOUR; if you missed it, check it out while you are here. This is a fine double star that is challenging for the smaller telescopes at medium power and a beautiful sight in all telescopes
Object 8 - Nice ACID TEST DOUBLE STAR for an 8-inch - Eta Corona Borealis
Although sometimes wide enough in its short 42-year period to resolve in the 4-inch, right now the companion to Eta Cor. Bor. is due east from the main star (mag. 5.7) only by about 0.4" arc, a true acid test for an 8" scope. You MUST use very high magnification and only attempt on the very darkest and most still nights to resolve this star....otherwise you will be disappointed.

Object 9 - Galaxy - The "Black Eye Galaxy" Messier 64 (Coma Berenices)
This one is a must-see in the 4-inch scope range and larger telescopes. However, small telescope users:

it is a great object for you as well, one of the brightest of all the galaxies visible in our scopes! At a full 8th magnitude and measuring 7.5' x 4' diameter, this is an incredible object. In the smaller scopes it appears as a very nicely compact oval; on a very dark night with about 100x, you might glimpse a little hint of the dark dust lane for which this galaxy is famous. It clearly bisects the bright hub of the galaxy, separating what you will see of this oval into about a 1/3 portion below and 2/3 above the dark lane. This effect is particularly pronounced and visible in the larger telescope, and can be glimpsed on very dark nights with higher magnifications with the smaller telescopes. High powers are recommended for the best viewing go M-64. The nature of this interesting "black eye" is a huge lane of dark, light-obscuring dust in the galaxy. At a whopping 44 million light years distant, this object is one of the finest of all deep sky objects for smaller telescopes.

Object 10 - Galaxy - Messier 85
At only magnitude 10.5 to the eye, this is a difficult object (but can be seen under very dark conditions) for small telescopes; it is featureless, even with

medium-high power and very dark skies in both the 6- and 8-inch, appearing as a uniformly-illuminated elongated oval with no distinct markings within. However, the larger scopes CAN see a distinct brightening as you progress toward the center portion of this oval. T hose with the 8-inch and larger telescope can probably spot another spiral galaxy (only about magnitude 10.7), ngc 4394 about 8' arc to the east of M-85, easily in the same field of view at medium magnification.

Objects 11 - Galaxy Group in Coma Berenices: Messiers 98, 99, 100
These three galaxies form a very large triangle very near to one-another in the sky; however, except with very low magnifications and a wide field of view, they likely will NOT be seen in the same field. All are faint magnitude 10.3 to 10.7 and thus out of the reach of the smaller telescopes. Even with 8" aperture, expect nothing but faint elongated patches of light, and very difficult at that. Use your GO TO under "messier object" to locate each of the individually. Of the three, M-98's shape is conspicuous as it is a near-edge-on spiral, very large (8' long) and appears like a pencil of light under dark sky conditions. M-99 has a very concentrated core or nucleus which can just barely be discerned in the 4-inch.

Object 12 - Galaxy - Messier 88
At magnitude 10.5, this galaxy is still a good object for all of our telescopes; it is one of the brightest of the large concentration of galaxies in this general area, and will handle magnification very well. Using averted vision with larger telescopes, some faint wispy detail might be glimpsed in this very

white and otherwise uniform image. It appears fairly large and round in all scopes, and should be observed with about 20x per inch aperture for best results.

Object 13 - Fantastic Edge-On Galaxy - NGC4565
This is the one you've seen in numerous photographs, the "needle" of light, the most famous of all edge-on galaxies. I have clearly seen the central dark dust streak for almost the entire 12' arc length of this thin streak of light in an 8" Schmidt on numerous times; under very dark conditions, the 4-inch shows it clearly against the enlarged "hub" of the galaxy's center. For all scopes, this is an amazing sight, appearing as a thin sliver of light that will fill a medium-power eyepiece of a good 3" telescope! In very dark skies, this is a remarkable object and its shape is plainly obvious even at low power. Its concentrated light makes for an almost-three-dimensional. In the larger telescopes, if you are fortunate enough to have high quality eyepieces, this galaxy provides an incredible sight and an unforgettable image for you.

NGC 4565 – ASO 0.51m Astrograph

Object 14 - Globular Cluster - **Messier 3** (in Canes Venatici)
Now, we're getting into some remarkably

memorable images! Messier 3 will be partially resolved even in a 3-inch telescope, and will show some "granulation" of stars in small APO refractors; however, in the 6-inch the stars really begin to come out at about 160x; in the 8-inch with a quality eyepiece, this globular is magnificent and even the star COLORS are evident on dark nights.

Globular clusters "take" power very well; I have found that about 230x is the limit for the 6- and 8-inch, and 120x for smaller scopes. A power of around 75-80x will do the best job on these clusters with small refractors.

The stars in M-3 are numbering in the 100's of thousands, and from earth all are 11th magnitude and fainter. Hundreds of these can be glimpsed on a very dark night at 160x with a 6-inch, and many scores of peripheral stars (outlying) can be seen with a 4-inch. Stars all the way into the center, perhaps thousands, can be glimpsed like stardust with larger scopes. In ALL scopes look for "streams" of stars that appear to radiate outward in noticeable strings from the center. This effect can be seen even if you do not fully resolve the cluster. Below is a visual drawing showing what might be seen in a modest telescope under dark conditions; notice the concentration of stars in the center of this cluster. The following illustration depicts the actual visual view that might be afforded under dark conditions with a 6 to 8-inch telescope.

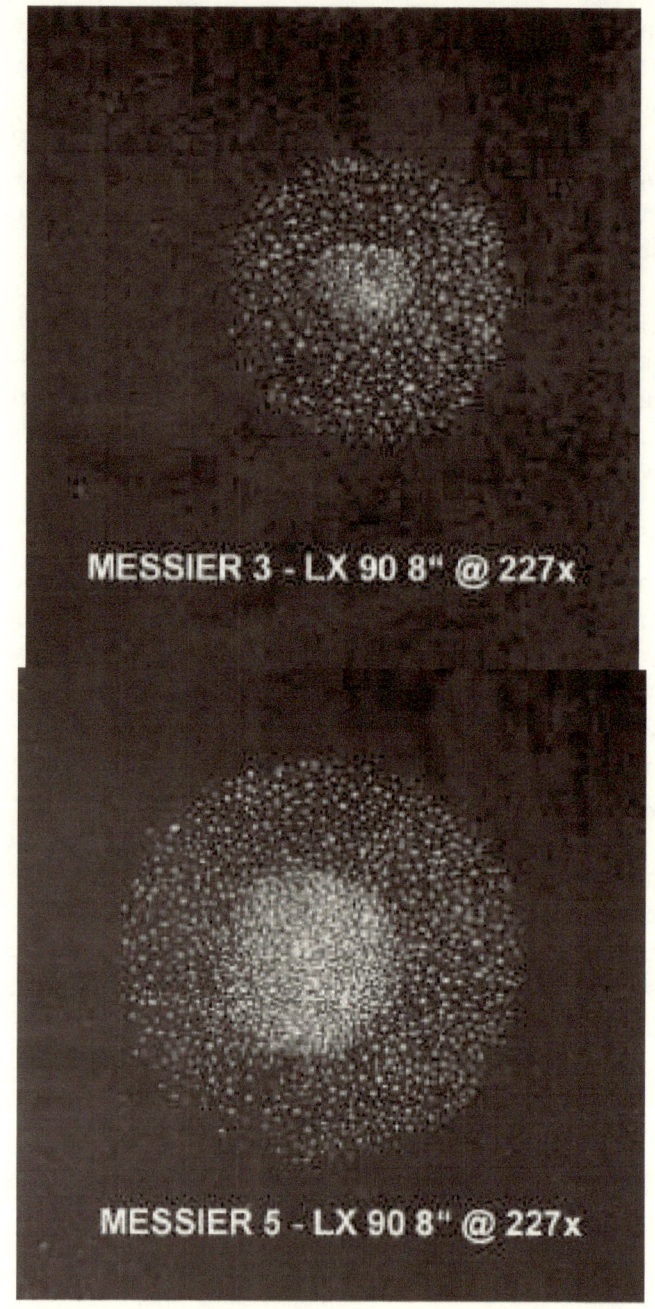

MESSIER 3 - LX 90 8" @ 227x

MESSIER 5 - LX 90 8" @ 227x

<u>Object 5</u> - Large Globular Cluster in Serpens - Messier 5

Clearly the **overall finest** globular star cluster of the skies; not appearing quite as large as well-known Messier 13 in Hercules because it is more distant, this cluster is one of the showcase deep sky objects.

In small and medium amateur telescopes you will see a brighter image of stars uniformly across this globular than with the sparsely (by comparison) M-13. The stars appear to outnumber those of M-13 as well, and range from about 11th magnitude to 15th; like M-5, a 4-inch will show peripheral stars along the fringes of this beautiful site with a "granulated" appearing and very bright core. Hundreds of stars across the entire object will be seen on a dark night with a 6-inch, and the globular is truly spectacular in an 8-inch, although I have yet to see distinguishable star colors in this one as I can clearly seen in M-5 and M-13. Even good binoculars (and your finderscopes) will show M-5 as a fuzzy ball in the sky and small scopes will give a spectacular low-power panoramic view of this object against a beautiful background of stars. For your information, M-5 is one of the oldest objects known in our galaxy, perhaps dating back 15 BILLION YEARS.

WANDERING ABOUT....YOUR NEW "USER OBJECT" FOR BOOTES

This brief GO TO tour of Bootes and nearby constellations has revealed their most interesting objects. But please do not stop here. Go ahead and locate many of the hundreds of other ngc galaxies that are present in this constellation.....there are several more interesting double and multiple stars that are within the reach of your telescope. In addition, use this opportunity to actually LOAD ANOTHER USER OBJECT onto your AutoStar, NexStar or computer planetarium program!

HOW ABOUT THIS FOR A SHOWCASE 'USER OBJECT?' We all know how to find "celestial north" or the South Celestial Pole here on earth. We rather take it for granted. Those poles are the location in space toward which our EARTH's axis of rotation points.

But what of the Milky Way Galaxy in which our Earth and its Sun are located? It spins on its axis, just like all galaxies.....where is ITS "**north galactic pole**" pointing? Here's YOUR chance to actually aim your telescope at the GALAXY'S NORTH CELESTIAL POLE! This is the point in deep space toward which our 250 billion-sun Milky Way galaxy's axis of rotation is pointing....just like Earth's north axis pointing toward Polaris! It just so happens that the "NGP" is located just inside of the constellation of Coma Berenices!

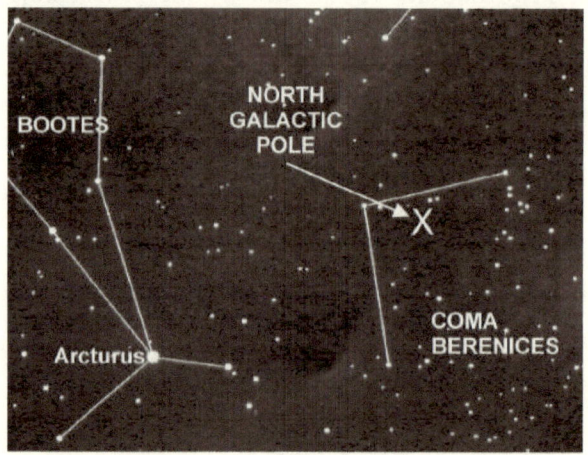

On AutoStar, go to: "Select/Object [enter] and scroll down to "user object" [enter]. Now enter these coordinates: **R.A. 12h 50' / DEC +27 52'** ; under "description" write something creative like "dead center" or something.

Please note that all sky programs, Apps and scope keypads have this same capability but each will require slightly different keystrokes. Using your program actively through such study guides as Constellations, you will quickly master exactly how to enter your own personal user objects, and become proficient at finding objects by entering only sky coordinates!

You will now have as your THIRD TOUR USER OBJECT the North Galactic Pole programmed to properly hold your guests in awe over.

Good observing and explorations of the wonderful world of deep space!

*   *   *

"Don't wait for the stars to align, reach up and rearrange them the way you want...create your own constellation...."

Singer/songwriter Pharrell Williams

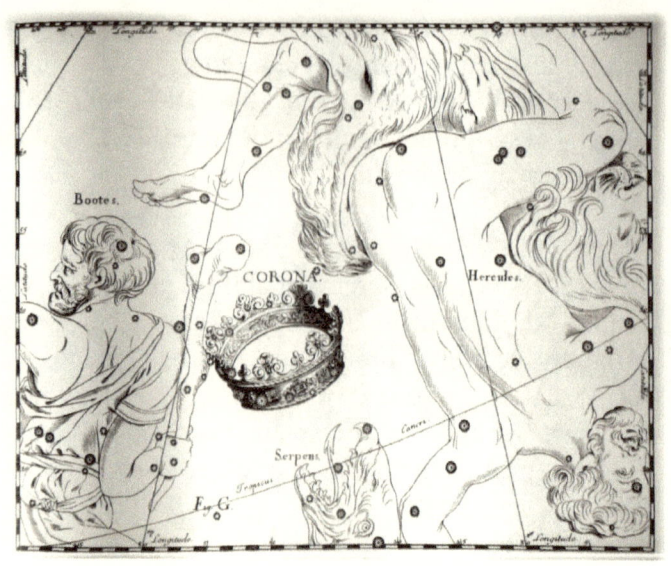

A celestial crown: Corona Borealis
From Hevelius' *Atlas* 1687

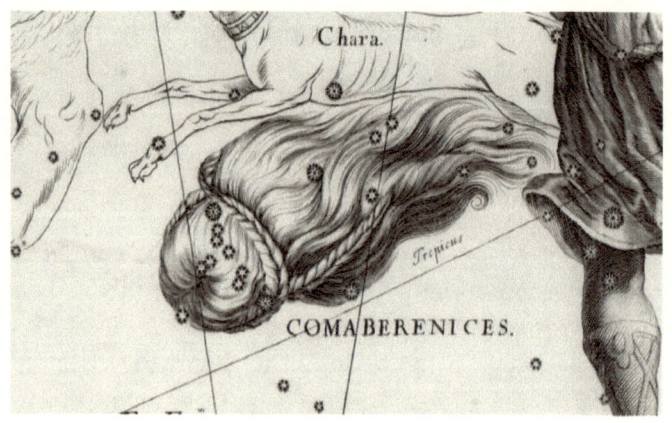

The tresses of Berenices – Coma Berenices
From Hevelius' *Atlas* 1687

Chapter Seven

CAMELOPARDALIS - a celestial "Giraffe"
....say what?? Is it a Giraffe....or is it a Camel?
What's in a name.....Let's take a look:

This is our next Constellation Guide, "GO TO CAMELOPARDALIS" of the series for computerized telescopes. So let's get the obvious out of the way early.....how many of you have actually ever even LOOKED at Camelopardalis? How many can actually point out where it is in the sky??

Not many, I'll bet, and when you look at the list of the brightest stars of this faint but large constellation, you can clearly see WHY you have not visited the giraffe pen any more often.

However, you have been missing out on some fantastic deep sky objects and some wonderful multiple stars if you have ignored the constellation as you will quickly realize from our chosen listing of interesting or unusual objects following

Camelopardalis is NOT one of **Ptolemy's** original 48 constellations, but a very faint group of "lost stars" that was stranded somewhere in between Cassiopeia to the west and Ursa Major to the east. Immediately NORTH of this sprawling stellar "wasteland" is the north celestial pole and the brighter star Polaris. It is one of the most recent additions to constellations, added by the German astronomer Bartsch.

Wherever his source might have been Bartsch

recognized this star pattern (after two millennia of being neglected from Ptolemy's famous four-dozen) as the "*camel*" of Biblical fame that was responsible for safely bringing Isaac from across the vast middle east to his awaiting Rebecca. Thus, the name is somewhat "close" to representing that story concept....but the actual translation of the word "Camelopardalis" to modern scripts reflects a *GIRAFFE*, not camel. Hence, earliest star tables such as the 1690 star chart by **HEVELIUS** (shown below) show something more of the long-necked creature.

Camelopardalis
as drawn by Hevelius in 1690
(Sherrod's "Llamadehors-elis")

Now...I ask the question: does Hevelius' drawing look anything LIKE a "giraffe?" We know the answer to that is: "..other than the long neck, NO." So...does it look like a "camel?" Clearly the answer to that is "NO." What it actually appears to be (at least to these old star-strained eyes) is a cross between a Llama and a horse...sort-of a *"Llamadehors-elis"*.

Another interesting note about this figure from Hevelius' old star catalog is his nature of rendering ALL star patterns and associated creatures/figures such as this. Compare (if possible) his drawing of key stars to those stars shown in my star chart above. They are REVERSED! All of Hevelius's charts were so drawn to represent how the star patterns would look if you were LOOKING BACK toward the EARTH, actually through the stars and toward the Earth. Very interesting....that would, indeed be quite difficult of today's modern GO TO telescopes, would it not?!

This constellation is another one of the **CIRCUMPOLAR CONSTELLATIONS** of northern skies, just as is Cassiopeia, featured in this series of "GO TO" constellation tours. This means that - from mid-northern latitudes of about 35 degrees and higher MOST of the stars of this large constellation neither rise nor set below any horizon, only spin around in apparent circular paths centered on the north celestial pole.

Camelopardalis is highest in the sky on midnight, November 15 and lowest (closest to the northern horizon, between Polaris and the horizon) on May 15. It is a very difficult constellation pattern to recognize, even through wide field photographs, and much imagination must have been used by **Bartsch** when he so designated this faint collection of stars to be either a Biblical "camel" or another animal NOT mentioned anywhere in the scriptures: a "giraffe."

To the naked eye, this constellation might appear at first a bit uninteresting, as there are no very bright stars nor conspicuous asterisms (star patterns) on which to focus your attention; nonetheless, there is an absolute WEALTH of nice deep sky objects (two nice planetary nebulae, by the way!) and many, many great double stars, only a few of which can be covered here. For a very complete listing with abbreviated details concerning nearly 100 multiple stars in this constellation, turn to my favorite guide: *BURNHAM'S CELESTIAL HANDBOOK*, Vol. One in which you will find the constellation listed alphabetically with the comprehensive multiple star listing at the beginning of the chapter on Camelopardalis.

As with every "GO TO" guide, each GO TO object in Camelopardalis is discussed for your telescope regarding the type of conditions necessary for you to view it optimally for discern the very faintest details.........magnifications and aperture necessary for most objects, and much, much more. This is YOUR complete guide to get you on your way to exploring the best (and few!) objects in this constellation. The following listing of "BEST" objects contains the finest or most interesting from my own observing experience and preference.

Use the attached star chart and the following Guide as an excellent reference for your next star party itinerary, or a beginning for further study into the thousands of objects visible in this part of the sky. Truly these extensive Constellation Study Guides will most definitely put your telescope to work for you in the most efficient and enjoyable way possible! As a matter of fact, MANY PC star program users are now programming their own "Tours" based on these guides, using each constellation as a separate GO TO Tour for the sky program library that can be added in or deleted through the main edit screen on your PC or MAC computer.

We hope you enjoy these comprehensive guides to touring the constellations via your AutoStar and other computer-driven telescopes. Each new installment is complete with diagrams, charts and illustrations that you will find nowhere else. Please let us hear YOUR feedback and your observations of each and every constellation after YOU have toured its vast reaches of our skies!

YOUR CAMELOPARDALIS CONCISE DIRECTORY OF INTERESTING OBJECTS –

I have chosen the finest (or most interesting) 11 objects in this CAMELOPARDALIS "GO TO" tour; as with all guides, all objects listed below will be visible in most telescopes (NONE naked eye in this installment, as there are no deep sky objects in Camelopardalis visible even in a pair of 7 x 50 binoculars....a fact that is also quite unique to this faint and largely-ignored constellation) Larger apertures may "show" an object a bit closer and "better," but frequently a wide field and low power view is more desirable than aperture for FINDING the objects initially. Indeed, I strongly encourage you first FIND the target object, or its approximate

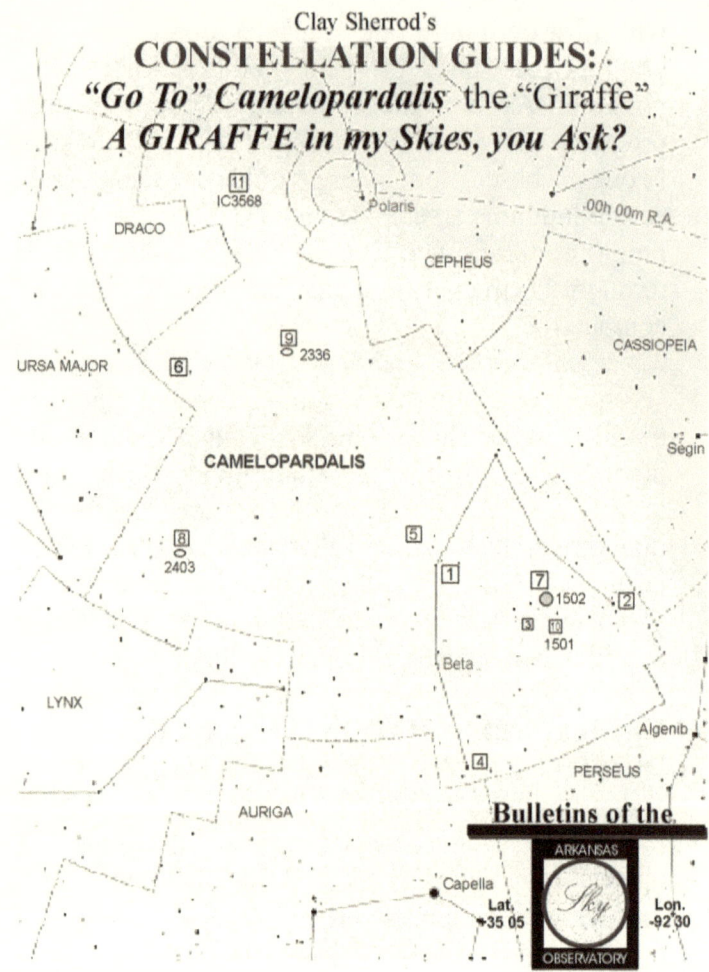

Clay Sherrod's
CONSTELLATION GUIDES:
"Go To" Camelopardalis the "Giraffe"
A GIRAFFE in my Skies, you Ask?

A finder chart for locating many of the GO TO
objects in the constellation of Camelopardalis; if
using a computer planetarium program, you are
encouraged to plot the objects on your screen for
higher resolution than this chart provides.

*   *   *

location through your GO TO function with your
lowest power and then - once IDENTIFIED

positively - move up slowly in steps with magnification if necessary. Remember, not all objects "like" magnification. Sometimes better "field of view" (such as with smaller telescopes) is desired over light gathering (like an 8-inch) and magnification.

The rule for determining "optimum magnification" is that: 1) too low power results in sky background glow detracting or diminishing the contrast against the deep sky object; 2) too high magnification darkens BOTH the sky background AND the object; 3) medium magnification can be achieved at which you have MAXIMUM contrast between the object and its darkened background sky. I have found through three decades of direct observing that about 15x per inch aperture for deep sky observing is PERFECT for most objects. That being said, always remember that DOUBLE or multiple stars require whatever power you can crank out....the seeing conditions are the limiting factor here.

For my complete and comprehensive discussion regarding seeing conditions and sky transparency, see also my GUIDE on that subject here at ASO .

With all deep sky objects, avoid attempting to observe when the moon is in the sky, even a very thin crescent, as its brightness in the sky will overshadow the very dim contrast afforded by even the brightest deep sky object; if you see the object at all against moonlight, you will NOT see the subtle outlying areas or the full detail of what is presented.

As with all of the "GO TO" tour constellation lists, I recommend a good star atlas and/or chart which will list all the finest objects, constellation-by-constellation. One very handy reference guide is the *PETERSON FIELD GUIDE TO THE STARS AND PLANETS*, which features complete lists with declinations, right ascensions, magnitudes, and all pertinent information for you to expand your observing horizons beyond this brief guide.

Use the attached star chart (above) and the following Guide as an excellent reference for your next star party itinerary, or a beginning for further study into the thousands of objects visible in this part of the sky. The chart gives the outline of the major constellation as well as the approximate locations of all objects discussed in this guide.

Following is the concise object list for your "GO TO" tour of CAMELOPARDALIS; you may wish to find the majority of the objects from the Library (for example, you can easily go to the nice planetary nebula ngc1501 if you pull up "Object/Deep Sky/NGC/..then type in '1501'...." and then press "Enter", followed by "GO TO" to access this object. On the other hand, if you want to experiment and become a "better computer user" try entering the exact R.A. and DEC coordinates of that object as described above after holding down the MODE key. You will find the accuracy of entered GO TO's to be somewhat less than those stored in the program, but the capability of acquiring unlisted objects is fantastic!

You will access your FIRST GOTO target - (usually the brightest star in each constellation) - different that we normally do in our "GO TO" tours, since Camelopardalis has NO "named" star! Instead, we will go to "ALPHA CAM" via the Right Ascension and Declination that you will enter just as described above! That will be a first for you, in all likelihood. If you don't first access this 4th magnitude star via entering its coordinates as provided in the following listing.....KEEP TRYING!!

You may also access the constellation by: SETUP / OBJECT / CONSTELLATION / Camelopardalis... Enter....GO TO.

OBJECT 1: bright star - alpha Camelopardalis ("Alpha Cam") - R.A. 04h 49' / DEC + 66 16' - Magnitude: 4.4

OBJECT 2: nice bright double - 2 Cam (Struve 385) - R.A. 03h 25' / DEC + 59 46 - Mags. 4.5 & 9 - A "fun" test

OBJECT 3: triple star - Kuiper 16 - R.A. 04h 19' / DEC + 59 30 - Mags: 6, 10.5, 8.1 - A wonderful object!!

OBJECT 4: test triple - 7 Cam - R.A. 04h 53' / DEC + 53 40 - Mags: 4.5, 8, 11 - challenge for a 5" scope!!

OBJECT 5: super variable - S Cam - R.A. 05h 36' / DEC + 68 46 - Mag. 8.1 to 11.0, 326 days!

OBJECT 6: very erratic star! - Z Cam - R.A. 08h 20' / DEC + 73 17 - Mag 10.1 to 13.0....a WILD ride!!

OBJECT 7: nice galactic cluster - ngc1502 - R.A. 04h 03' / DEC + 62 11 - Mag. 5.3! 15 stars, VERY small!

OBJECT 8: nice spiral galaxy - ngc2403 - R.A. 07h 32' / DEC + 65 43 - Mag: 8.8, VERY large....nice!

OBJECT 9: nice spiral - ngc2336 - R.A. 07h 16' / DEC + 80 20 - Mag: 11.0 - spiral arms can be seen!

OBJECT 10: faint planetary nebula - ngc1501 - R.A. 04h 03' / DEC + 60 47 - Mag: 13.3 w/ 13th mag. star!

OBJECT 11: another planetary - IC3568 - R.A. 12h 34' / DEC + 82 51 - Mag. 11.6 w/12th mag. star. Near Polaris!!

A VISUAL GUIDE TO OUR DEEP SKY OBJECTS IN CAMELOPARDALIS

Object 1 - Our "Starting" Star – alpha Camelopardalis
You will have to use your knowledge of you sky program to find this one! Alpha Cam is an un-named 4.4 magnitude star that you likely would NEVER look at if it were not for this "TO TO" TOUR! Enter the R.A. and DEC coordinates above as described early in this Guide into the program and press "GO TO" to zero in on this rather uninteresting and uneventful star. It is NOT double....it is NOT variable...it does NOT have an distinctively pretty color....it IS, however, a very "early" type "O" star, very hot and bright white at a distance of some 1,000 parsecs from our solar system, more than 10 times more distant than any other brighter star in Camelopardalis!

Object 2 - Double Star "2 Camelopardalis" (also known as Struve 385)
There is a bit of history in this star as it was one of the first to be cataloged as "double" by Sir William Herschel in the 18th century with his "40-foot Reflector" (see my article about this scope at the Arkansas Sky Observatories website:
www.arksky.org

This is a very nice for small and larger telescopes, with a clear separation of about 2.3" arc. The primary star is a relatively bright magnitude 4.5 while the companion star (nearly due south from the brighter one) is a faint magnitude 9.1. Both stars are blue-white, and were re-cataloged by W. Struve early in the 19th century. With the difference in brightness, this is a nice star for all scopes, since it is NOT as easy as it first may look!

Object 3 - Kuiper 16 - A Fine Triple Star
Here is a tough test for even the 6-inch and certainly anything smaller; although the two main components ("A" and "B") are a full 1.3" arc and should theoretically be resolvable through a good 3" scope, I have found this a particularly difficult pair to cleanly split because of a great difference in brightness. The first two stars (1.3" arc apart) are magnitudes 6.2 and 10.3 (!), with the fainter star seemingly VERY close nearly DUE SOUTH of the 6th magnitude star, and just a little bit west. The third component - a true gravitational triple - is found much farther away and is MUCH easier to see! Look for this "C" component nearly NORTH EAST from the brightest star "A", a whopping 32" arc away....that is slightly less than the entire width that Jupiter would appear in the same eyepiece. To see the tough pair ("A" and "B"), use very high magnification in the 3" and larger scopes (about 40x per inch aperture minimum); the distant "C" component will be much easier (you can spot it with an 80mm) using medium power (about 20x per inch aperture). All three stars are "early spectral type-A" stars, very brilliant yellow-white.

<u>Object 4</u> - Another Triple Star! 7 Cam - Provides Quite a Test for the 5" Telescope!

Here is a triple star that will be a challenge for all scopes 5" and larger. This is a true triple star that is moving through space with a common proper motion. The odds are stacked against you to see all three components, but it CAN be resolved into all three components in a 5" scope theoretically; I have never been able to see all three stars in any 6" Newtonian telescope, but have cleanly resolved all in a good 6" refractor on steady nights using about 300x. In all respects, it should not be a problem with an 8" or larger scope, but still an interesting challenge! Here is the problem with this star.....see if YOU can overcome the odds! (refer to my chart following:

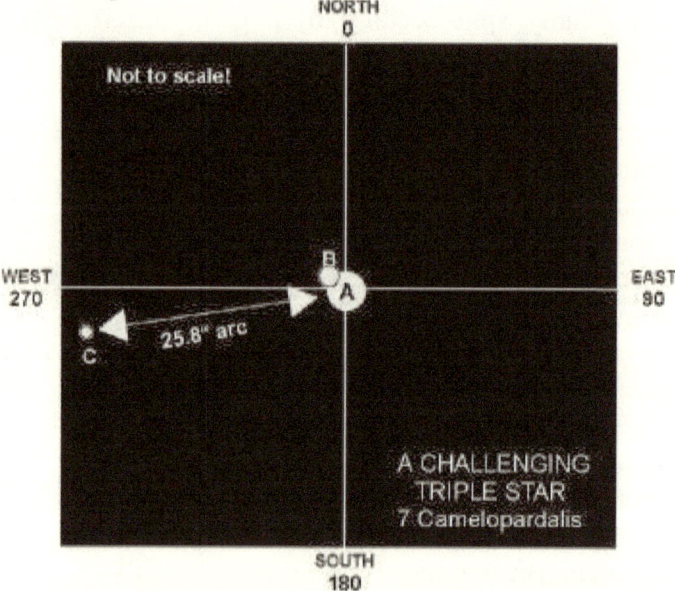

1) Components "A" (primary) and "B" are separated ONLY by 0.9" arc, the exact theoretical resolution limit of a perfect 5" aperture telescope. However, it

gets harder. There is a huge magnitude difference when you add that to the proximity of these first two stars. "A" is magnitude 4.4 and "B" is magnitude 8.1, thus the brighter star tends to overshadow the secondary "B" star. Look for the fainter of these two in Position Angle 285 degrees, or nearly due WEST of the brighter star.

2) Component "C" adds insult to injury....it is far enough away to be easily seen a 4-inch and larger (26" arc - about the size of Mars' disk in the same eyepiece) BUT it is magnitude 11.2! So carefully look for this fainter companion star to the other two in Position Angle 240 degrees (just south of due west) fairly far from the primary, nearly in the same angle as the "B" star is offset.

Object 5 - A Very Nice "Year-long" Variable Star - S Camelopardalis
To observe this star, you will need two star charts from the American Association of Variable Star Observers (AAVSO - www.aavso.org) , the "finder" chart - a low power, wide field chart to help you locate the star the first few times out as well as make magnitude estimates when it is at its brightest - known as "a chart"; this chart can be downloaded from:
https://www.aavso.org/apps/vsp/ . Note for these charts, simply type in the NAME of the variable at top to generate your choice of chart –
next, for observing the star when faintest you will need a narrow field, faint-comparison-star chart ("g chart") providing stars below 11th magnitude. Note that both of these AAVSO charts can be reversed as you will see the field in a catadioptic telescope through a diagonal mirror.

On all AAVSO charts, they are free for the downloading and you merely save them to file on your computer, call up the file once done and resize (they are huge when you download!) and save again in an appropriate size; then merely print them out for use at the telescope!

S Camelopardalis is a Long Period Variable, ("LPV" - see my complete discussion on variable star observing and star types in the Observing Variable Stars GUIDE at ASO Guides on this website) but is an interesting one in that it is not really all that predictable as to its light curve....see the chart below:

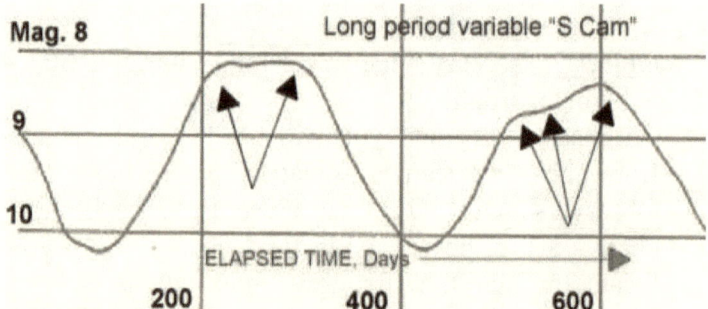

Most LPV's have a maximum peak that is fairly brief and quick to "round out" once achieving maximum and begin immediately fading again; the LPV stars are huge and very red pulsating giant stars whose pressure pushes the outer layers away from center (brighter) and eventually this expansion is overcome by gravity and the star's gasses are once again pulled back inward again (dimmer). Note that with S Cam, the typical maximum brightness, magnitude 8.1, is consistent, but it sometimes will remain there for up to 60 days....other times, such as the SECOND peak you see in the light curve above, there are minor fluctuations while at brightest. The MINIMA (around magnitude 10.8) is always very consistent and short-lived. This is an excellent star for all scopes, large and small. It can be seen throughout its 326-day cycle in even a 2" refractor, but good magnitude estimates (which are invited for submission the AAVSO, by the way!) should be done with somewhat larger instrument of high quality. Observe this star about once every two weeks to log YOUR estimate for an annual light curve of S Camelopardalis!

Object 6 - An Erratic "*Dwarf Nova*" – Z Camelopardalis
Here is a wild star as can be seen from the nearly 2-year light curve below:

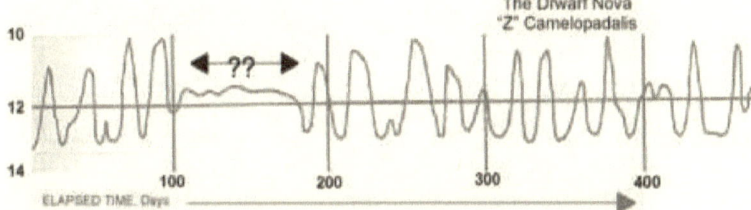

This is a very cataclysmic star, erupting about every 2 to 3 weeks, but never so regularly that its next event can be accurately predicted! As can be seen, its rise to maximum - nearly 3 full magnitudes! - takes place in just over ONE DAY, as does its rapid demise once peaking out. The maximum nor minimum never lasts more than a couple of days at most. The star is a dwarf binary star system it is thought, with a cooler star so close that it gradually expels volatile gases from gravitational pull from a more massive and hotter star nearby....when these gases reach the outer layers of the hot star, the flare up violently and result in the remarkable brightening we see.

The range for this star keeps it well within reach of 6-inch and larger scopes, but it is a pretty difficult star to monitor regularly with the 3" scope, although it CAN be done. The maximum is typically a very consistent 10.1 or 10.3, while the minimum is always about the same - 13th magnitude. It is the period of outbursts of this star that are NOT predictable, and sharing YOUR estimates of this star's light changes is certainly encouraged by the AAVSO; unlike S Cam., this star should be observed at EVERY opportunity! It can go from a barely-seen speck of light to one of the brighter stars in your medium power field of view!

To locate the star and to observe it with comparison stars when it is at brightest, log on the AAVSO site again and download the "a" chart: https://www.aavso.org/apps/vsp/ . Note for these charts, simply type in the NAME of the variable at top to generate your choice of chart.
When the star is at dimmest, your will need the "g" chart. NOTE: unlike the charts for S Cam above, these charts are newer "Reversed" charts ("BR" and "DR" in those two URL's listed)....these are charts that the AAVSO has designed with today's modern telescope user in mind....the field of view MATCHES that of your catadioptic; in other words, North is at Top and East is always to the Left on the charts, just as in your telescope!

Object 7 - NGC 1502 - A Very Small, Compact Galactic Cluster with Two Double Stars!

This is a very irregular galactic cluster of loosely-associated stars. With a total magnitude of a seemingly bright 5.3, it is clearly visible in all scopes, but as a very small object compared with other larger star clusters. It is only 7' arc across and there are between 15 and 25 stars in a wide range of magnitudes associated with it. It is interesting to me because the star field surrounding ngc1502 is SO VOID of stars that this tiny cluster shows up distinctly like a small compact pile of diamonds on black velvet. This effect can clearly be seen in the ASO photo below:

Note from this photo the wide range of star magnitudes. Indeed, two very easy and enjoyable double stars (actually one is a quadruple star!) are found within this cluster:

STRUVE 484 at: R.A. 04h 03' / DEC +62 12 (mags. 9.1 & 9.4, 5.3" arc apart - great in a 3-inch scope!) and,

STRUVE 485 at: R.A. 04h 03' / DEC +62 12
(mags. 6, 6.1, 12.5 and 13!)

Both of these multiple star systems will be evident
at medium to medium-high magnifications when
you have zeroed in on the ngc1502. Most of the
stars in this cluster are magnitude 8, but there are
several brighter ones seen in the photograph; thus
Struve 485, with its two primary stars (mag. 6 and
18" arc separation) should be clearly distinguishable
in the cluster as one of the brighter stars....once
found, look for Struve 484 immediately next to it,
but as a much closer pair.

<u>Object 8</u> - A Nice Bright Galaxy - VERY Large –
NGC 2403

Nice and Large Spiral Galaxy
NGC 2403

Size can be deceiving, particularly with "extended"
celestial objects, those whose brightness is spread
out over a very large area. For example the

Andromeda galaxy is a bright magnitude 3, but that is what we call "total Integrated Magnitude", or the entire brightness of the object if every bit of it was compressed from its large size (nearly 2 degrees for M-31!) into the size of a star.

Thus, the large galaxy ngc2403 shown above in the U.S. Naval Observatory photograph has a total seemingly bright magnitude of 8.8 spread out over a field of view half the size of the full moon's apparent diameter - 17' arc! Nonetheless, this is a galaxy worth looking at...on a very dark night, use a nice wide field eyepiece at very low magnification to best see this object. Note the large "clumps" in the photograph which are large clusters of dense star groups in this spiral. With a 6-inch and larger scopes these can be clearly seen with a magnification of about 100x. In 8" and larger telescopes this is a fine, and detail-filled object! In a small APO and other wide field instruments - even binoculars! - look for this beautiful object as a large "glow" about 8 million light years away....some four times farther than the Andromeda galaxy.

Object 9 - NGC 2336 - very difficult except in 8"
A "many-armed" spiral galaxy!
This is a very difficult object, magnitude 11.0 visually, and fairly small (5.7' long by 2.8' wide). I am including it because this is a very worthy spiral galaxy for observation with 8, 10 and 12-inch telescopes and we have many readers in that size range. It can be seen distinctly in a 5-6 inch, but I cannot make out anything but a "fuzzy smudge", very faint and set in a fairly nice field of stars. However, in VERY dark conditions, observers with the 8" and larger telescope might glimpse

unmistakable multiple arms of this galaxy, resembling as a very faint pattern of what we familiarly see in photographs of the famous "Pinwheel Galaxy."

Object 10 - Planetary Nebula, NGC 1501 - Nice Test for 5" Scope

This is not an easy object, but once found, it gets better the longer your eyes become acclimated to the surroundings. This round planetary nebula (remnants of a massive stellar explosion) is about the apparent size of Jupiter as seen in a medium-power eyepiece. Use about 200x to best view this bluish-green disk-like object, and always observe only when the skies are very dark and without any moonlight. There is a central star (the one that created the nebula) of magnitude 13.4, but the nebula will obscure sighting of this star in the 5" scope, while the nebula itself should be clearly visible. In the 8" and larger telescopes, the central star is seen in fleeting glimpses, coming and going as conditions warrant. It is worth waiting for when that tiny speck of light that created this object "jumps out" and suddenly sparkles in your eye!

Although the central star for this planetary nebula - a shell of stellar gases emitted by the explosion of an unstable star - is bright enough for a 6-inch scope limiting magnitude, it is VERY difficult to see visually except in larger telescopes; the reason is that there is much extraneous gases that block the visible wavelengths of light to the eye; averted vision in an 8-inch will reveal the star but only with a magnification of up to 400x and the darkest of nights. This star is plainly visible and appears erroneously "large" in the Palomar 200" photograph shown above.

Object 11 - Our Final Object - Planetary Nebula IC 3568
Here is another nice planetary nebula, one that is a bit brighter than the previous object. IC 3568 is about half the distance as ngc1501 but appears only 1/4 the size! (12" arc diameter). This very small green disk can be seen in the 5-6 scopes and

perhaps only a bit better in the larger 8" scopes. However, it IS brighter (magnitude 11.6) and thus will hold increased magnification better than ngc1501. The central star of this planetary is ALSO visible as a magnitude 11.9 object and thus should be within reach of the 5" scope on a very steady dark night...normally an 11th magnitude star would be easily discernible with this scope, but the nebula - as explained above - obscures it for the most part. A trick to observing this - and all other central stars of planetary nebulae (try this on the Dumbbell, M-27, and the Ring, M-57!) is to use a good quality "nebula filter" which really isolates the nebula from the surrounding sky and stars....the central star will "pop out" at you!

WANDERING ABOUT....YOUR NEW "USER OBJECT" IN CAMELOPARDALIS

We are going back to actually the BRIGHTEST star in Camelopardalis for our new "USER OBJECT" to input into the Autostar for this "GO TO" tour. No....we are not reverting back to Alpha Camelopardalis with which we started our tour, but to Beta Cam, which is actually the brightest of all objects in this constellation! At magnitude 4.2, it outshines Alpha by 0.2 magnitudes, so Bayer had his brightness classifications mixed up a bit, or his eyes were more blue-sensitive since Beta is a bit of an orange colored star.

Beta Cam (that sounds like an obsolete video device, doesn't it!?) is actually a very nice and easy triple star, even for smaller instruments as can be seen in my double star diagram below.

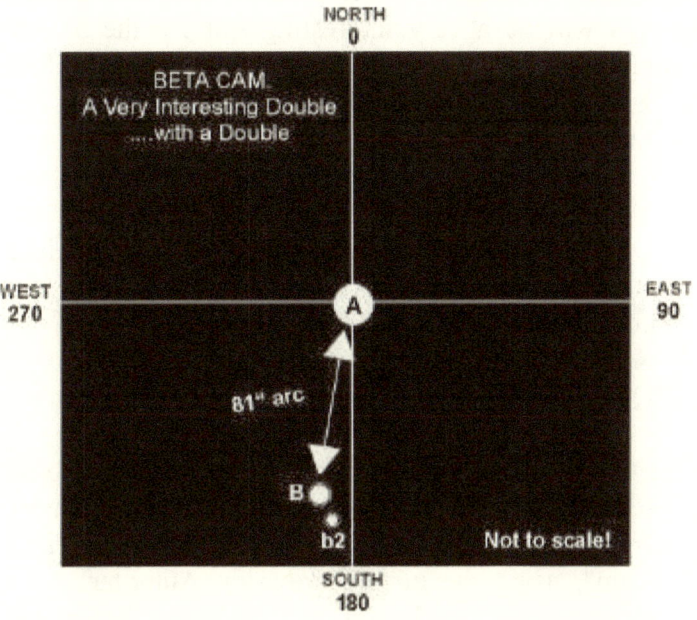

NORTH
0

BETA CAM.
A Very Interesting Double
....with a Double

WEST
270

EAST
90

A

81" arc

B

b2

Not to scale!

SOUTH
180

The primary star "A" is magnitude 4.2; the "B" companion is nearly due south in Position Angle 208 degrees from "A" a whopping 81" arc seconds distant. That spacing would require two Jupiters in your eyepiece to fill, so being too close is no excuse to NOT see this 9th magnitude star; in smaller telescopes, I recommend at least 75x to increase the image scale enough to see what will be a faint "B" star. In larger telescopes, use lower magnifications for best views. Now...the "b2" star companion is actually a companion to the "B" star....NOT the "A" star, so that is what makes this an unusual system! Look for "b2" as an 11th magnitude faint star (should be visible in a 4-inch, certainly so in the 5" and larger instruments) 15" arc also fairly south of the "B" star....not from "A" remember, but from the 9th magnitude companion to it!

Quite a challenge just keeping up with it! Certainly this will be a very interesting star for the 5" and larger scopes and likely for 3" telescopes as well. It is worth looking at, and worth remembering the facts.

Your will need to GO TO this star via its R.A. and DEC coordinates, just as we did for the Alpha starting star for this "GO TO" TOUR. The coordinates for Beta Cam are:
R.A. - 00 hours 59m
DEC. - + 60 degrees 22m

On AutoStar or similar, go to: "Select/Object [enter]...." scroll down to "User Object" [enter]. Now enter the coordinates given above for "beta Cam", using the number keys on AutoStar. After entering the coordinates and pressing "Enter" yet again, scroll down one and you can list the magnitude of the object as "4"[Enter].

This will provide you with yet another unique deep sky object for conversation-starters and crowd-stoppers at your next star party or club outing....NOT a true "triple star" but a double star with a third star that strings along!

*   *   *

"My thoughts are stars that I cannot fathom into constellations...."

John Green

Chapter Eight

CAPRICORN

In case the "Dolphin" and "Little Horse" did not fit the zoological bill....how about a "Sea Goat?"

Welcome to our Constellation Guide for, "GO TO CAPRICORN" of the series "GO TO guides for all GO TO Telescope Users." Far in the southern skies (high overhead for many of our regular readers hovering around the Earth's equator!) in the midst of the periphery of the watery "ocean" of the universe, just east of the great Milky Way's "celestial river" of the ancient Egyptian cultures is a wandering "Sea Goat" who represents one of the twelve constellations of the Zodiac.

Although most of us are familiar with the constellation name and its significance astrologically (there's an oxymoron for you), few have ventured into is expansive boundaries in search of celestial wonders. There is a good reason for that...

....there aren't any.

Nowhere within the constellation will you find the likes of the Orion Nebula, the Andromeda Galaxy, the Pleiades, star clouds, Ring Nebulae, globular clusters. This is a constellation that is noticed by its absence of notable and, frankly "interesting" deep sky objects. So other than the one nice globular cluster Messier 30 in Capricorn, this installment of your Constellation "GO TO" TOURS will concentrate on observing the many (and there are, indeed MANY!) fine double and multiple stars within the brighter stars of this constellation.

IN ADDITION - there is a nice bonus and purpose for your visiting Capricorn: the distant planets URANUS and NEPTUNE are perched near there for the next many years. Since these planet move VERY slowly against the background stars because of their extreme distance from our viewing station here on Earth, they will be in this constellation for many years. Check your sky program coordinates for weekly, monthly and yearly changes in position from those provided below (epoch June 10, 2010).

Note that this fairly inconspicuous constellation is in a relatively star-poor area of the sky sandwiched between the fabulous object-rich areas of Sagittarius (to its west) and an equally star-poor Aquarius to the east. Even the areas nearest the Celestial Equator are 10 degree south (-10 degrees declination) so ALL objects in Capricorn will have negative (-) declinations in the lists below.

The *ECLIPTIC* - this being the 10th of the Zodiacal constellations - passes right through the entire span E-W of Capricorn, so it is often than planets will visit its realm. As a matter of fact.....for the next few years, as stated you can enjoy TWO planets within Capricorn: **URANUS** and **NEPTUNE**, so I recommend that you refer to your program for the latest locations. You will find some brief but practical information regarding observations of these planets below.

Also appearing in this installment is a quick "observing tips summary" for proper nomenclature,

CONSTELLATION GUIDES:
"Go To" CAPRICORN"
Taking a Look at the Garden Variety "Sea Goat"

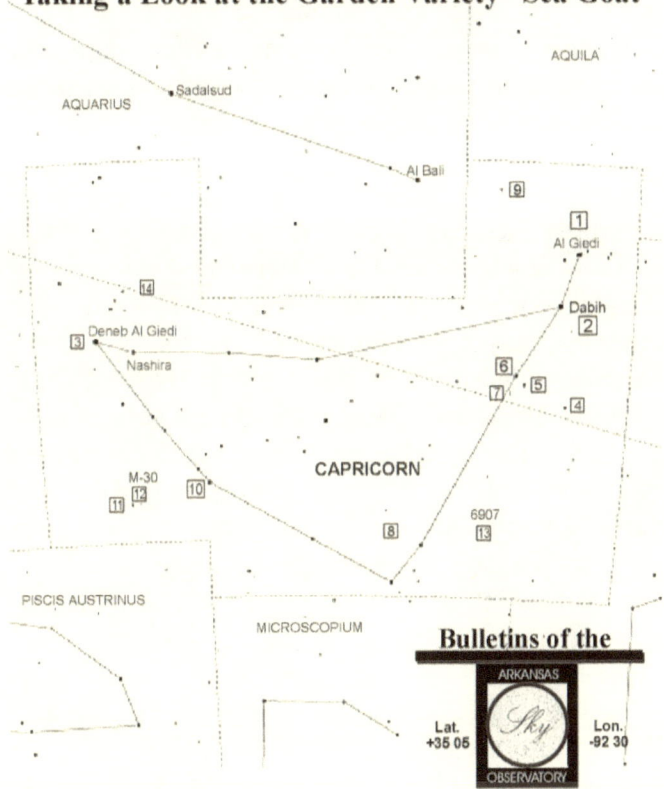

A finder chart for locating many of the GO TO objects in the constellation of Capricorn; if using a computer planetarium program, you are encouraged to plot the objects on your screen for higher resolution than this chart provides.

*   *   *

positioning and observing techniques for double stars. There are several stars noted here that will make excellent test objects for specific models of our GO TO telescopes, so make note of them if you have been wanting a challenging test for the overall optical performance of your telescope and eyepiece combinations.

Each GO TO object is discussed for your telescope regarding the type of conditions necessary for you to view it optimally for discern the very faintest details....double star challenges for each size telescope .....magnifications and aperture necessary for most objects, and much, much more. This is YOUR complete guide to get you on your way to exploring this large constellation. Although some interesting and "test" double stars as well as on nice variable star is included in the "GO TO" tour, there is a particularly wide range of very interesting double and multiple stars too numerous to list or mention here.

After our usual Concise List of objects with RA and DEC coordinates for your reference, there will be a complete and abbreviated listing of good double and multiple stars for most telescopes from 60mm to 200mm in size. In addition and for more information about these stars consult a good handbook, such as the "*Burnham's Celestial Handbook*," Vol. 1 for a very comprehensive list of locations, magnitudes and angular separations of these wonderful stars.

Use the attached star chart (above) and the following Guide as an excellent reference for your next star party itinerary, or a beginning for further study into the thousands of objects visible in this part of the sky. Truly these extensive Constellation Study Guides will most definitely put your computer to work for you in the most efficient and enjoyable way possible! As a matter of fact, MANY telescope users are now programming their own "Tours" based on these guides, using each constellation as a separate GO TO Tour for the AutoStar or PC sky program library that can be added in or deleted through the main edit screen on your PC or MAC computer.

We hope you enjoy these comprehensive GUIDES to touring the constellations via your PC sky program and its computer-driven telescope. Each new installment is complete with diagrams, charts and illustrations that you will find nowhere else. Please let us hear YOUR feedback and your observations of each and every constellation after YOU have toured its vast reaches of our skies!

YOUR CAPRICORNUS CONCISE DIRECTORY OF INTERESTING OBJECTS –

There is a listing found at the end of our "concise list" of selected "best" objects that gives an abbreviated description, position, magnitude and size of each of the fine multiple stars found in Capricorn. I urge you to challenge yourself and equipment to seek out these stars....explore strange new worlds...go where no man...., (no, better stop there).

ALSO remember to "go to" both **Uranus** and **Neptune** as part of this tour, as described as your last two objects in this TOUR. Note that the coordinates for both planets will VERY gradually change from those given, but not by much. Your sky program has the wonderful capability of calculated the exact position for each day, hour and minute that you choose to observe these planets from ANY particular observing location.

For your specific "GO TO" tour I have chosen the finest 14 objects in Capricorn (and, boy, was it hard to find some!), mostly double stars...two deep sky objects and one variable star; as with all GUIDES, all objects listed below will be visible in all telescopes (some naked eye) from a 3-inch to 8-inch telescope; of course larger apertures may "show" an object a bit closer and "better," but frequently a wide field and low power view is more desirable than aperture.

NOTE: unlike deep sky objects, the many double stars listed here can actually be observed very well in **lighted suburban skies**, provided that your telescope can reach the magnitude specified for the faintest star of the group you are observing. Steady air is the most important aspect in good double star observing, NOT sky transparency (see my GUIDE on this subject under the ASO GUIDES tab on the Arkansas Sky Observatories website) as is required for deep sky observing of faint, extended objects.

The very low altitude (-10 to -30 degrees declination) of Capricorn in our mid-to-late summer skies lends itself well to very good and long-period observing for telescope users in the most southern states of the U.S. and Europe, and ideally suited for observers between +20 degrees and -20 degrees. When rising about dark in the east in early summer,

Capricorn will remain in the sky throughout the night, transiting the meridian at midnight about August 9. All objects and difficult double stars are ALWAYS best observed when they are located nearly overhead (or as high in the sky as possible), thus requiring the observer to look through the thinnest portion of the Earth's "lens" of atmosphere and haze.

As with all of the "GO TO" tour constellation lists, I recommend a good star atlas and/or chart which will list all the finest objects, constellation-by-constellation. One very handy reference guide is the *PETERSON FIELD GUIDE TO THE STARS AND PLANETS*, which features complete lists with declinations, right ascensions, magnitudes, and all pertinent information for you to expand your observing horizons beyond this brief guide. For the computer, many references are available by simply doing a search for "Capricorn".

The constellation tour Star Chart shown previously will get you started on your journey for this constellation.

Following is the concise object list for your "GO TO" tour of Capricorn; NOTE: most of these double stars listed, and those particularly in the abbreviated list following are NOT listed among your computer library for "STARS / DOUBLE". For these you MUST enter the coordinates of RA and DEC for them and let Autostar find the positions for you. To do this, merely press down and HOLD the "Mode" key for about three seconds; scroll down to the display showing your RA and DEC for the current position in which your telescope is oriented;

For others, you may wish to find some of the objects from the telescope library (for example, you can easily go to "Messier 30" if you pull up "Object/Deep Sky/Messier Object/..type in '30'...." and then press "Enter", followed by "GO TO" to access my favorite beautiful galactic cluster. On the other hand, if you want to experiment and become a "better computer user" try entering the exact R.A. and DEC coordinates of that object as described above after holding down the MODE key. You will find the accuracy of entered GO TO's to be somewhat better than those stored in a handbox, and the capability of acquiring unlisted objects is fantastic!

FOR PLANETS (Uranus and Neptune in this case), go to "SELECT /OBJECT/SOLAR SYSTEM...." and use your lower left scroll key to move until finding each planet by name; press "ENTER", let it calculate its position for you, then "GO TO" at which time the telescope will take off and center the tiny - BUT VISIBLE - planets for you for the EXACT date you are out, no matter when you use this tour!

OBJECT 1: your Capricorn index stars (2) - Al Giedi (alpha Cap) - R.A. 20h 15' / DEC (-)12 40 - **Magnitude: 4.5, 3.8**
OBJECT 2: nice triple star! - Dabih (beta Capricorni) - R.A. 20h 18' / DEC (-)14 56 - Magnitudes: 3.0 & 6.0 & 10
OBJECT 3 history! - Deneb Al Giedi - R.A. 21h 44' / DEC (-)16 21 - Magnitude: 2.8 - Where Neptune was discovered

OBJECT 4: another double star - Sigma Cap. - R.A. 20h 17' / DEC (-)19 17 - Magnitudes: 5.5 & 10, wide nice double

OBJECT 5: great double star! - Pi Capricorni - R.A. 20h 25' / DEC (-)18 22 - Magnitudes: 5 & 8.5 - wide, pretty!

OBJECT 6: tough LX 90 test double - Rho Cap - R.A.20h 26' / DEC (-)17 59 - Mags: 5 & 9 - only 0.5" separation!

OBJECT 7: great double star! - Omicron Cap. - R.A. 20h 27' / DEC (-)18 45 - Magnitudes 6 & 6.5, great for smaller APO refractor!

OBJECT 8: 4-inch test double - Burnham153 - R.A. 20h 44' / DEC (-)26 36 - Mags: 7 & 8 - 1.6" arc, great test!

OBJECT 9: nice double - Burnham668 - R.A. 20h 30' / DEC (-)10 02 - Mags: 6 & 11 - close, nice in 4-inch & above

OBJECT 10: 6-inch test double - Zeta Cap. - R.A. 21h 24' / DEC (-)22 38 - Mags: 4 & 13 - wide, but a challenge!

OBJECT 11: double in nice view - 41 Cap. - R.A. 21h 39' / DEC (-)23 29 - Mags: 5.5 & 12 - great star for small telescope test!

OBJECT 12: nice globular cluster!! - Messier 30 (ngc7099) - R.A. 21h 38' / DEC (-)23 25 - Mag. 8.4, nice, but small

OBJECT 13: small galaxy - ngc6907 - R.A. 20h 22' / DEC (-)24 58 - Mag.12 , very faint, but visible in 6- to 8-inch

OBJECT 14: variable star - UU Cap - R.A. 21h 34' / DEC (-)14 06 - Mag. 8.7 to 11, semi regular, very interesting!

OBJECT 15: URANUS - (use "Solar System" GO TO) - Mag: 5.7 / Size: 4.4" arc / Color: Blue Green

OBJECT 16: NEPTUNE - (use "Solar System" GO TO) - Mag. 7.8 / Size: 2.2" arc / Color: Yellow-blue

OBJECTS 17 THROUGH 41 –

MULTIPLE/ DOUBLE STARS in Capricorn (other than those listed above)

Listing of nice multiple stars in order of RIGHT ASCENSION visible and resolvable most telescopes. NOTE: These doubles are NOT detailed in the following "Visual Guide" -use the abbreviated descriptions ONLY stars of interest in 2.5" through 8" aperture telescopes are provided.

Use the following standard nomenclature for proper star label:

h = Herschel
B = Burnham
numbered star = Aiken ADS listing
E = Struve STAR SYSTEM
R.A. and DEC.
type
magnitudes, directions & separations (" of arc)
SCOPE LIMIT (aperture in inches)

MULTIPLE STARS IN CAPRICORN

h156 - RA 20 04 / DEC -09 04 / triple, mags 7 (primary), 10 (east - 6.2"), 12 (NE - 8.4") - 3"
3 - RA 20 14 / DEC -12 30 / double, mags. 5.5, 13 (NNE - 27") - 6"
E2683 - RA 20 26 / DEC -13 18 / double, mags 9, 9 (ENE - 23") - Good test for 6"
h2975 - RA 20 31 / DEC -22 24 / double, mags. 7.5 & 11 (N - 9.8") - excellent for 3"+
h1537 - RA 20 34 / DEC -15 29 / double, mags.

8.5 & 8.5 (N - 3.3") - good 3" test object! NICE

h40 - RA 20 41 / -19 40 / double, mags. 8.5 & 9 (NE - 5.2") - great test object for 2.5"!

B674 - RA 20 42 / -21 04 / double, mags. 8 & 10.5 (E - 1.6") - perfect test for 4"

h2998 - RA 20 46 / -20 48 / double, mags. 8.5 & 9 (ESE - 5.9") - possible in all scopes

h5226 - RA 20 47 / -27 33 / double, mags. 7 & 8.5 (E - 19") - good in 3"

h3003 - RA 20 50 / -23 58 / double, mags. 6 & 8 (S - 1.7") EXCELLENT for 4" and larger!

B271 - RA 21 17 / -26 33 / triple, mags. 7 (primary), 9.5 (W - 3.2"), 12 (E - 82") - great object for 5"+

B1262 - RA 21 20 / -15 07 / double, mags. 8.5 & 9 (ESE - 2.0") - 3" test star

B683 - RA 21 25 / -20 26 / double, mags. 8.5 & 11 (S - 2.9") - another good one for 3"

Aiken2096 - RA 21 31 / -16 25 / double, mags. 7 & 10.5 (NE - 0.8") - super test for 8"!

B168 - RA 21 51 / -20 15 / double, mags. 8 & 9.5 (SE - 5.7" arc) - use high power in 3"!

h3071 - RA 2155 / -15 23 / double, mags. 7 & 11 (NW - 18") - great star for the 3"

....AND NOW ON WITH THE SHOW!! (refer to the Star Chart for all the objects described in detail on the "Guide")

A VISUAL GUIDE TO OUR DEEP SKY OBJECTS IN CAPRICORN

Object 1 - Our Starting Star in Capricorn - "Al Giedi" (alpha Capricorni) - Also a COMPLEX MULTIPLE!!
For the record, this is known as the "goat star" from

the Arabic in association with the odd association as a "sea goat" for this constellation. For a change, this is a REALLY interesting and nice starting point for our "GO TO" TOUR. Alpha Capricorni is NOT the brightest star in this constellation, that honor being taken by DELTA ("Deneb Al Giedi" discussed below). But this is no ordinary star. With a 6 arc' separation, Al Giedi (NOT "Deneb Al Giedi") is a wonderful double star for the small APO refractors (3-inch), with the brightest (alpha 2) being magnitude 3.5 and the other (alpha 1) being magnitude 4.2. Both stars are wonderful in a wide field ETX 60/70. Now, here is the good part: BOTH stars have companion stars as well! So this double is also a double (same scenario as the more brilliant and famous "double-double" star in the constellation "Lyra" (see *Constellations* Guide Vol. II / **Lyra**). Except in this case, the two brighter stars ARE not a physical pair, only an OPTICAL double, two stars appearing close but are really far apart in space. Alpha 2 is only 100 light years from us, while Alpha 1 is over 500 light years distant.

Alpha 1 and Alpha appear exactly 6' arc apart, with the fainter star being about due WEST of Alpha 2. Alpha 2 (brighter star) has a small companion star at 11th magnitude some 6.6" arc almost due SOUTH from the bright star, so it should be relatively easy to find (although faint) in a 4-inch scope. Alpha 1 (fainter star) actually has TWO companions: the first is magnitude 9, about 45" arc (the size of Jupiter in the same eyepiece) just WEST of SOUTH from the brighter star; due SOUTH of the brighter star is another companion at magnitude 13 the same distance exactly. Alpha 1's 9th magnitude companion is an easy target for a 3-4 inch, but the 13th magnitude star will require a 6" or larger for spotting. I have made up the following chart (oriented correctly for a compound scope with NORTH at top and EAST at right) to assist you in the proper orientation of these five stars.

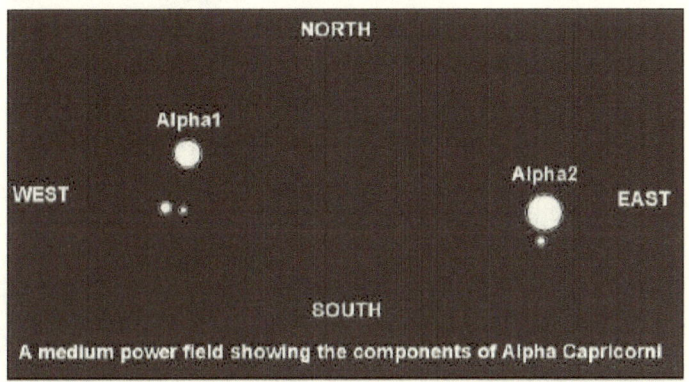

A medium power field showing the components of Alpha Capricorni

Object 2 - Beta Capricorni - Another Good Double - nice color contrast, great for a small telescope!

At about the same distance as Alpha 2, above, Beta Capricorni ("Dabih") is 150 light years distant and given a visual magnitude of right at 3.0. The primary star is magnitude 3.0 and the secondary star is magnitude 6.1, about 3 arc minutes DUE WEST of the brighter star. This makes an easy object for ALL telescopes at fairly low power. However, the secondary star is ALSO a double, this having a faint 10th magnitude companion DUE EAST of it, ONLY 0.8" arc. This means that only an 8- or 10-inch is capable of resolving this star, and then only on the very best nights; about 227x is required for clean separation. The small star will appear like a "knob" on the "right side" of the 6th magnitude star.

Object 3 - Delta Capricorni - "Deneb Al Giedi" - A Historical Marker in the Sky!

Delta Capricorni, a 2.8 magnitude star, is the brightest star in the constellation and is very close to Earth, only 50 light years distant. There is not much out of the ordinary about this star other than the fact that in 1846, the LAST planet that was discovered "off" American soil - **Neptune** - was found only 4 degrees northeast of this relatively bright star! What makes this even MORE interesting at the present time is that over 150 years later, Neptune is BACK VERY close to where it was originally found after its 165 year trip around the sun. So, you can see where history was made....and be part of history when Neptune returns to the place of its "conception" in 2010! (for observing Neptune, see the description following).

Object 4 - Sigma Capricorni - A Great Double Star for All Telescopes - A Challenge for a good 3-inch telescope.

Here is a nice example of a very "late" spectral type "K" double star; the primary star, Sigma Cap appears distinctly red at magnitude 4.1, and it has a 10th magnitude companion located almost exactly 1' arc DUE SOUTH (easy to find that way!) from it. If you can remember how "big" Jupiter looks in a medium power eyepiece (about 75x should do it), that is the space from the brighter star due south to the companion. This should be easy in a 6- or 8-inch, but worth the effort in a smaller telescope.... but certainly attainable, and a challenge (because of the faintness of the secondary star) in the 3" scopes.

Object 5 - A Wonderful Double Star for all Size Telescopes! - Pi Capricorni

This is a "classic" double star, and very bright at that. Overall, the two-star system combines for a magnitude of 5.1, with the brighter star ringing in at magnitude 5.2 and its companion at 8.5. Look for the fainter star (very easy in all scopes - use about 20x to 30x per inch aperture) only 3.2" arc away almost exactly SE from the main star. This is a very brilliant white-blue pair of stars. This is also Burnham's B60 on his famous double star list.

Object 6 - Rho Capricorni - A VERY Tough Triple, even for an 8-inch - Here is your test! 6-inch scope bonus!
Maybe you can do it....maybe you can't. I can split this double star with 300x in my 6" Unitron refractor, but have not had the opportunity to attempt it with a modern 8-inch. The 6-inch catadioptic will NOT even elongate this star on the best nights. Separation?? How about 0.5" arc?! That is a challenge. Rho Cap is a fifth magnitude star, easy to find through its coordinates. The companion star SHOULD be visible even in smaller instruments, at magnitude 9, but it is so close that it is nearly impossible to differentiate. HOWEVER, Rho gives us a BONUS star....a 13th magnitude true companion some 55" arc exactly SE of the brighter star. THAT one can be see with a 5-6 inch on a very dark night when the constellation is highest in the sky.

Object 7 - Omicron Capricorni - Super Double for ALL Telescopes!
We have gotten one really tough double (but, we all like a challenge!) out of the way, so let's focus in on a really pretty and easy double star. Omicron Capricorni is a wonderful yellow, type-G double

consisting of a 6th magnitude primary star with a 6.5 magnitude companion just south of WEST from the brighter star. Since the separation is a whopping 22" arc (half the size of Jupiter in the same eyepiece), even a good 60mm APO refractor should show this double with medium to medium-high magnification; in a 3-inch and larger scopes, use very low power (the 26mm eyepiece is a good choice) to appreciate the real beauty of these two nearly equal brightness stars among a very star-poor background.

Object 8 - Another Double - A Great Small Telescope Star....one of **S.W. Burnham's** First: #B153
This is a super test star for the ETX 90. If you think your optical system is well collimated, cleaned and ready, scope out this one. B153 is a pair of 7.5 and 8.5 magnitude stars, very sun-typical. At only 1.6" arc separation this will make an ideal test for the 3.5" scope! Theoretically, the ETX 90 should knock a hole right in between these two stars....but sometimes life's tough. Use very high power to achieve a clean split between these two stars. Look for the fainter star DUE WEST of the brighter component. In a 6-inch the stars will be cleanly split but very close together at 227x.

Objects 9 - A Very Nice Classic Double - Ideal for the 3-5 inch scope range - Burnham #668
B668 is a very nice test star for a good 3-inch, wider in separation (a full 2.9" arc) but with a much fainter (11th magnitude) secondary star. The primary star is magnitude 6, so it will be clearly visible in the finderscope and easily identified in a low power field of view in all telescopes; higher

magnification (about 175x and above) will be necessary to provide enough contrast with the 90 to see the faint star, which is NE of the primary. Look VERY close and attempt to block out the brighter star with a crosshair from a reticle eyepiece if possible! This is a relatively easy object in the 6- and 8-inch telescopes.

Object 10 - Zeta Capricorni - A Striking Double Star, a Test for the 5-6 inch scopes and some 8-inch Here is a marked contrast in brightness, the primary star being 4th magnitude and the secondary ONLY 13th magnitude! Zeta Cap's components have shown NO change in position of the small star relative to the brighter one since its discovery as a variable. Due to its brightness, there is no mistaking the star, but move up in magnification to better than 200x in both a 6-inch and 8-inch to actually "find" this faint star, which is nearly DUE NORTH of Zeta., but a full 21" arc - a good spacing - from the bright star; again, a crosshair blocking Zeta's light will GREATLY assist in your acquiring a view of the 13th magnitude companion.

Objects 11 - A Double Star With a View - 41 Capricorni - Wonderful Star for the 4 inch and Larger Scopes
This wonderful double star makes quite a sight in a medium power eyepiece, with the globular cluster Messier 30 (see below) just WEST of this star; if you can find Messier 30, then you have found 41 Capricornus at a bright magnitude of 5.5, easily visible in the finderscope and low power eyepiece. The photograph below shows the globular M-30 in relation to 41 Capricorni to its west (left)....note that in catadioptic scopes, the star will appear RIGHT of

M-30, or a mirror image of this astronomically-oriented photograph.

41 Capricorni is a 5.5 magnitude star of "medium" G-type spectrum. Very close (about 5.5" arc) to its SW is a 12th magnitude star, which should present a challenge in a 3-inch telescope, right at the visual limit of that scope. The star should be seen steadily in a 6-inch on a dark night, and relatively easy with the 8inch scope.

Object 12 - Globular Cluster Messier 30 (while you're in the neighborhood!)
Finally! A Deep Sky Object in a seemingly vacuum of black space! This is the lone globular cluster that is accessible to visual telescopes in the entire large cluster of Capricorn! Nonetheless, it is an object worthy of observing! Messier 30 is rated at magnitude 8.4 visually but brighter photographically at nearly 6.2. M-30 is best known for its "star strings," seeming lines of stars that

emanate from the central part of the globular. Speaking of the center, this one has one of the most tightly condensed central cores of all globulars, a feature clearly seen even with a 3-inch which will NOT be able to resolve even the peripheral parts into individual stars; some faint stars will be glimpsed on very dark nights with a 6-inch and a few more with the 8-inch with powers of about 150x. In all telescopes this relatively small globular will appear as an oval (not circular) glow, only about 4' long by 3' arc wide, with a very concentrated central brightness. This central core actually measures nearly 2' arc, almost disk-like. In the same field of view, using a wide angle eyepiece in all scopes, you will see the double star 41 Capricorni due east of M-30 (see previous photo).

Object 13 - A Lone Galaxy Amidst a Sea of Blackness - ngc6907
This is a very small and faint (magnitude 11.8) galaxy that MAY be visible on a very dark night in the 6-inch and can be glimpsed routinely with 8-inch scopes and larger. It measures only a small 2.5' x 2' in diameter and is a typical spiral galaxy (although NO detail is visible to the eye even in the largest telescopes). Fittingly, this isolated galaxy is located in one of the most "star-poor" regions of the summer sky, far in the southwestern "corner" of Capricorn.

Object 14 - Our "GO TO" TOUR Variable Star – UU Capricorni
Go to the American Association of Variable Star Observer's (AAVSO) wonderful web site for all the free star charts for serious variable star observing that you can possibly download! For UU

Capricorni, only a "d" (medium-high, small scale and faint comparison star) chart is available at: https://www.aavso.org/apps/vsp/ . Note for these charts, simply type in the NAME of the variable at top to generate your choice of chart. Merely download this chart to file, pull up and resize to fit your paper and print for a wonderful comparison chart at the telescope.

UU Capricorni is a great mid-range variable, semi-regular in its light fluctuations. That means that sudden and unexpected light changes can occur in a very short time frame, so constant observation of this star is always needed to "fill in the blanks." It varies in brightness from magnitude 8.7 to almost 11, making it an ideal star to watch for even a 3-inch and larger telescopes all the way through its 100-day cycle. This is a very red star (can be see as such at brightest) and a late "K" spectral type.

Object 15 - **THE PLANET URANUS** –
Since the planet Uranus was located in the constellation of Capricorn until 2003, (now in *PISCES* – see Volume II) this will be included as a "GO TO" object; consult your PC sky program for slight daily changes in its positions as it creeks slowly eastward through the Capricorni stars. In 2003, the planet will progress eastward and enter the boundaries of Aquarius, to the east where it will remain for some time.

Uranus has rested in the "sea goat's" territory now since 1996. Now at a distance of 1.811 billion miles from Earth, look for Uranus as a blue-green nicely defined disk in 3-inch and larger telescopes, at magnitude 5.8, easily definable in the finderscope.

With a small disk diameter of only 3.6" arc at the present time, don't expect to see very much here.

Uranus was discovered by **Sir William Herschel**, and documented as a new planet from its motion against the background stars in Taurus; as the years progressed, **J.E. Bode** confirmed Herschel's discovery as he tracked Uranus through Taurus and on into Gemini (the bottom foot in the ancient 1784 engraving seen below). Note that on August 1, Bode had sketched Uranus due north of the right foot of one of the Gemini twins! Chart reproduced courtesy Archives of the Arkansas Sky Observatory.

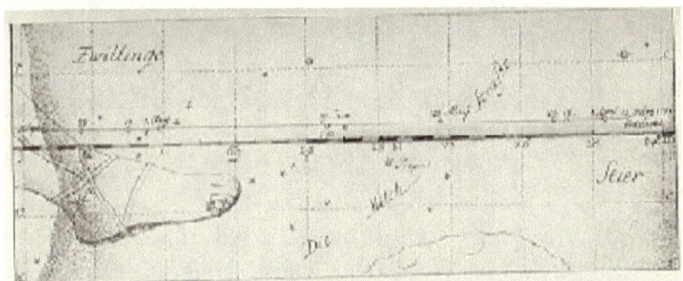

BODE'S CHART TRACKING THE MOTION OF URANUS
The "foot" of Gemini is at left.
Archives of the Arkansas Sky Observatory

The larger telescope and steady skies will show an occasional belt or "stripe" across the planet, similar to the gaseous cloud patterns seen on Jupiter (see my Observing Guide to Jupiter in ASO GUIDES on the Arkansas Sky Observatories website) as shown in the drawing below.

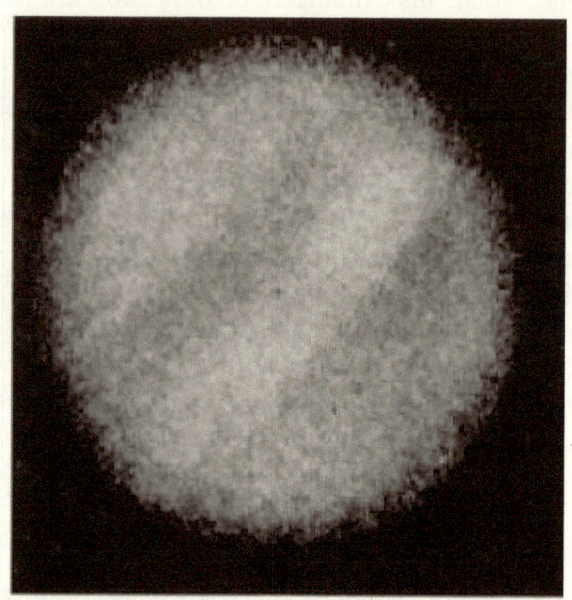

THE APPEARANCE OF URANUS IN THE LX 90
Arkansas Sky Observatory

In all telescopes, including the small ones, this distinguishable from the stars, that is bluish in color; the two main equatorial belts (see drawing above) will appear slightly lighter than the rest of the globe. There ARE occasionally white cloud spots visible in scopes such as the 8-12 inch range that can be followed and timed to accurately determine its rotational rates which are close to providing a "day" on Uranus of about 10 hours 49 minutes. However, while in Capricorn for the next few years, note that the E-W axis of Uranus will be up and down and the poles oriented left and right as clearly demonstrated in the beautiful color-enhanced Hubble Space Telescope photograph (courtesy NASA) shown below. Note also the rings which encicle the equatorial plane of the gaseous

216

planet.

The 6-inch scope might be able to spot three of the four major Uranian satellites (Ariel - mag. 13.7 / Titania - mag. 13.7 / and Oberon - mag. 13.8). An 8-inch will be challenged to pick up a fourth - "Umbriel" at magnitude 14.2; all will be see very close to the planet and as "stars" just on the limit of visibility. The following sketch by Herschel in February 1787, some time after the planet's discovery, shows the two brightest of these satellites.

A SKETCH BY SIR WILLIAM HERSCHEL
SHOWING THE TWO BRIGHTER MOONS OF URANUS

Arkansas Sky Observatory

<u>Object 16</u> - **THE PLANET NEPTUNE** –

As mentioned, the planet Neptune was actually discovered here in Capricron on <u>September 25, 1856</u>, almost 150 years ago. Presently it is in the constellation of *Aquarius*. Since its orbital period around the sun is 165 years, Nepturne is again re-entering the place of its discovery to mankind. Like Uranus, this tiny planet (disk is only 2.3" arc!) appears very bluish in the telescope at high power which is needed to differentiate it from surrounding faint stars. It is magnitude 7.9 presently and will be in Capricorn much longer this decade than will be Uranus. Use your Autostar planetary library to compute its daily eastward path through the constellation. Little or no detail can be expected on this distant planet at its incredible distance of 2.73 billion miles from Earth. Look carefully at the NASA photography of Neptune taken with the

Hubble Space Telescope (HST) and note the conspicuous darker feature that is very much like the Great Red Spot on Jupiter and the many bright white upper level clouds visible.

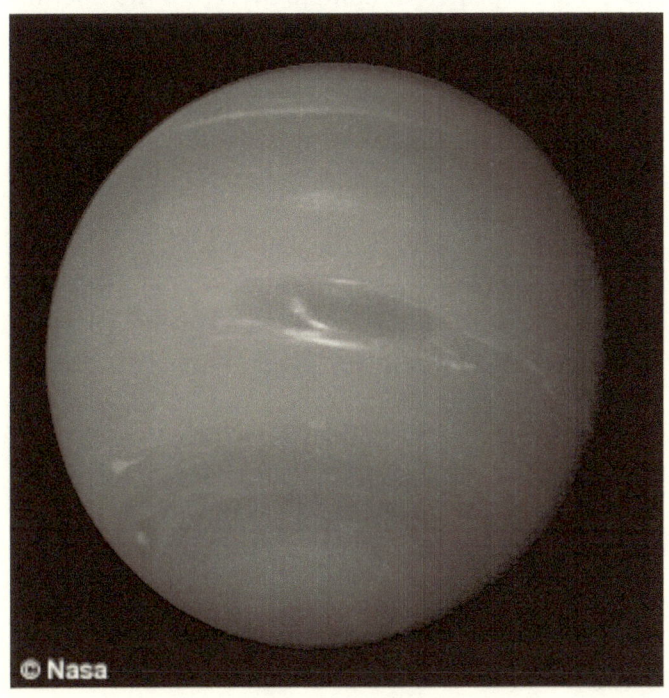

© Nasa

WANDERING ABOUT....YOUR NEW "USER OBJECT" IN CAPRICORN

This is an unusual entry for our sky computer User Object library: we are going to enter a "Historical Marker," the position at which **Sir William Herschel** discovered the planet Uranus on March 13, 1781 with a "7-foot" (focal length) telescope, thinking that his discovery was at first a bright and condensed COMET! Since it was found only 4 degrees NE of *Deneb Al Geidi*, this is a spot that will be interesting to all of your visitors at the next star party! Curiously I find myself looking at this spot often, noting that it is the LAST place that a human being "found" a planet visually. I look at those stars repeated to humble myself, remembering that 220 some years before me, one of the greatest astronomers who will ever live was peering at those SAME stars, yet there was an unexpected interloper among the....the planet Uranus.

And I sigh and reflect....

On AutoStar or PC program, go to: "Select/Object [enter]...." scroll down to "User Object" [enter]. Now enter the coordinates given above for "Deneb Al Giedi", using the number keys on AutoStar. After entering the coordinates and pressing "Enter" yet again, scroll down one and you can list the magnitude of the object as "7"[Enter]. Now go back and enter the NAME of this User object merely as: "History."

*   *   *

"Things are as they are. Looking out into it the universe at night, we make no comparisons between right and wrong stars, nor between well and badly arranged constellations ....."

Alan Watts

Chapter Nine

CASSIOPEIA
A Slightly Naughty and Precocious Celestial Queen
....some people see it as an "upside down W"....I see it as an "M"

In this Constellation Guide, "GO TO CASSIOPEIA" we will explore the pre-fall skies that serve as a precursor to impending crisp evening skies, the smells of browning leaves as Mother Nature prepares its trees for the crunch of winter cold.

Cassiopeia is one of the few *CIRCUMPOLAR CONSTELLATIONS* for the northern hemisphere in that it truly never rises nor sets below any horizon from as far south as 30 degrees north. Rather, it appears - if you were able to observe it throughout its 24-hour spin - to describe a complete circle centered on the North Star, Polaris. Note from my following diagram the positions of Cassiopeia as seen at 10 p.m. your local time on four times of year.

NOW....look at its changing positions throughout the course of a winter evening, from 6 p.m. through 6 a.m. local time:

(See following charts)

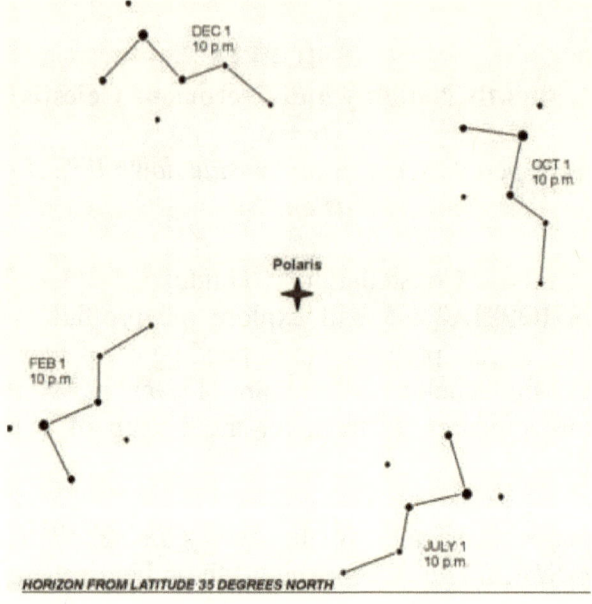

DEC 1
10 p.m.

OCT 1
10 p.m.

Polaris

FEB 1
10 p.m.

JULY 1
10 p.m.

HORIZON FROM LATITUDE 35 DEGREES NORTH

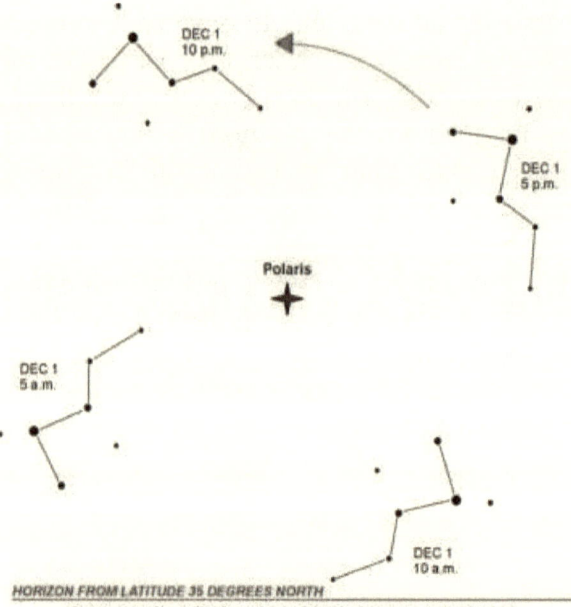

DEC 1
10 p.m.

DEC 1
5 p.m.

Polaris

DEC 1
5 a.m.

DEC 1
10 a.m.

HORIZON FROM LATITUDE 35 DEGREES NORTH

Circumpolar Chart #2 - Showing the Motion of Cassiopeia in 24 Hours

Through this constellation, perhaps even better than "the big dipper" in Ursa Major can we witness the rotation of the Earth through the course of the night as its geographical axis is pointing directly toward the central point of this great circle inscribed by the large "W" or "M" shaped asterism of the sky. Note that on midnight on or about April 10 each year, middle northern latitude observers will spot Cassiopeia very low on the northern horizon "below" Polaris; at this point the asterism describes a clear "W" shape; however, a half-year later we wind the same constellation also at midnight much higher in the sky "above" Polaris and turned "upside down" to form a distinct "M" shape on about October 10.

Cassiopeia is a familiar constellation to even the most casual skygazer, its primary bright stars (see below) describing that familiar "W" shape (Or an "M" is you are of that persuasion) in the high northern sky. The image below demonstrates the primary stars, marked with their common name designations in the constellation:

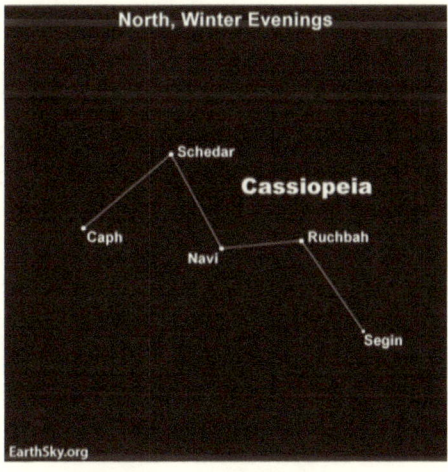

I have written much about the questionable carrying-ons of *Queen Cassiopeia* (wife of Cepheus) In Ethiopian, Arabic and Greek lore/ The daughter of Cassiopeia and *Cepheus* (see my Constellation Guide; Andromeda here on this ASO website) was the princess *Andromeda* who was left to be devoured by the great sea monster *CETUS* (*Constellation Guides: CETUS* under ASO GUIDES)by none other than mum and dad when their kingdom and subjects were about to fall victim to the wrath of the Gods of the Sea. . In addition to this dysfunctional family atrocity, the mother (Cassiopeia) was so vain that she bragged to all around her as more beautiful than the prized daughters of *Nereus*, one of the many gods of the sea.

Rumor has it - as told "Here First....only in the Enquirer...." that Cassiopeia was also unfaithful to her King husband Cepheus, philandering it seems with the muscular ("...it's not buttuh....") hero *Perseus* - although all the king's horses and all the king's men could never get their surveillance cameras to get the goods on the unfaithful spouse. So at least for her incessant vanity, Cassiopeia was destined to suffer the revenge of the god Poseidon by chaining her daughter, Andromeda to a rock a on a cragged rocky coast as a sacrifice for a sea monster, now enshrined as the constellation CETUS (now configured in more peaceful and "happy" images as a mere "whale"). But Andromeda was, alas, saved from certain "sea monster death" by the hero, *Perseus*. After rescuing her he demanded Andromeda as his wife (which Cepheus gladly accepted)...he came out a winner for both mother and daughter it seems! We have no late-breaking news as to the opinion of all this father-daughter bartering in the opinion of the lovely - but jilted - Cassiopeia.

So....however you might interpret this somewhat culturally conflicting story of the *Peyton Place* of the immortals, one fact about Cassiopeia remains intact in all cultural legends of her demise, either from

1) infidelity to Cepheus through the likes of the handsome "...it's not buttah"... Perseus; or,
2) her vanity that insulted all the gorgeous nymphs of the sea (known today as the "Swedish Bikini Team")

....the forlorn Queen has forever been banished to the sky encircle upside down forever, tethered to her once-royal throne. It is this "throne" that we see today as inscribed by the "W" shape (I can clearly see a "Shaker-style" four-post chair myself), and I have, for the likes of me, never been able to see that gorgeous figure of a woman we know as "Cassiopeia."

Cassiopeia (pronounced "cass-eo-PEA-a") is up all night for most observers in the northern hemisphere. It is at its "lowest" (closest to the north horizon) point at midnight on April 8 each year, with all of its size (6) brightest stars of the "chair" barely being seen as they spin under Polaris against the horizon at 35 degrees north latitude. Conversely, the conspicuous "W" shape is highest in the sky on about October 8 at midnight every year (midnight culmination). An interesting note about this circumpolar motion of Cassiopeia is that you will quickly realize that this conspicuous asterism of stars is ALWAYS nearly exactly opposite of the "little dipper" of the constellation Ursa Minor.

THE STARS OF CASSIOPEIA –

**NOTE: Listed below are the primary (all Greek alphabet ["Flamsteed"]) naked eye stars which comprise Cassiopeia. For reference, the Smithsonian Astrophysical Observatory (SAO) number is also provided, followed by the visual magnitude and spectral type of each star.

Bayer/Flamsteed   SAO   MAG   R.A. & DEC (2000)

Alpha 18 SHEDIR   21609   2.23 K0IIIa
00 40 30.465 + 56 32

beta 11 CAPH   21133   2.27 F2III   00
9 10.719 + 59 08

gamma 27 MARJ/Cih   11482   2.47 & 2.3

delta 7 RUCHBAH   2226   2.68 A5III
IV   01 25 48.986 + 60 14

epsilon 45 SEGIN   12031   3.38 B3III
01 54 23.71 + 63 40

zeta 17   21566   3.66 B2IV   00 36
58.297 + 53 53

eta 24 ACHIRD   21732   3.44 & 11.6

F9V+dM0   00   49   6.04   +   57   48

theta 33 MARFAK   22070   4.33
A7V   01 11 6.183 + 55 08

iota   12298   4.52 A5pSr   02 29
3.97 + 67 24

kappa 15   11256   4.16 B1Ia   00
32 59.965 + 62 55

lambda 14   21489   4.73 0.5 B8Vn
00 31 46.411 + 54 31

mu 30 MARFAK   22024   5.17
G5Vb   01 8 16.36 + 54 55

nu 25   21729   4.89 B9III   00 48
50.06 + 50 58

xi 19   21637   4.8 B2V   00 42
3.899 + 50 30

omicron 22   36620   4.54 B5IIIe   00 44
43.532 + 48 17

pi 20   36602   4.94 A5V   00 43
28.113 + 47 01

rho 7   35879   4.54 G20e   23 54
23.05 + 57 29

sigma 8   35947   4.88 & 2.8 B1V   23
59   0.532   +   55   45

tau 5   35763   4.87 K1IIIa   23 47

upsilon1 26   21832   4.83 K2III   00 55
0.125 + 58 58

upsilon2 28   21855   4.63 G8IIIbFe-0.5   00
56 39.82 + 59 10

phi 34   22191   4.98 133.8 F0Ia   01
20 4.908 + 58 13

chi 39   22397   4.71 G9IIIb   01 33
55.928 + 59 13

psi 36   11751   4.74 23.2 K0III   01 25
56.034 + 68 07

omega 46   12038   4.99 B8III   01
55 59.947 + 68 41

For a very comprehensive and complete listing and cross-reference of Bayer, Flamsteed, SAO, double and other star information for the constellation of Cassiopeia, I highly recommend that you visit: http://www.deepskywatch.com/deepsky-guide.html

GETTING STARTED –

As with every "GO TO" tour guide, each GO TO object in CASSIOPEIA is discussed for your telescope regarding the type of conditions necessary for you to view it optimally for discern the very faintest details.........magnifications and aperture necessary for most objects, and much, much more. This is YOUR complete guide to get you on your way to exploring the best (and few!) objects in this small constellation. The following Chart from the

Arkansas Sky Observatories and the subsequent detailed listing of "BEST" objects contains the finest or most interesting from my own observing experience and preference.

This finder chart for locating many of the GO TO objects in the constellation of Cassiopeia; if using a computer planetarium program, you are encouraged to plot the objects on your screen for higher resolution than this chart provides.

Clay Sherrod's
CONSTELLATION GUIDES:
"Go To" CASSIOPEIA
A Beautiful but Less-than-Ladylike Queen of the Sky

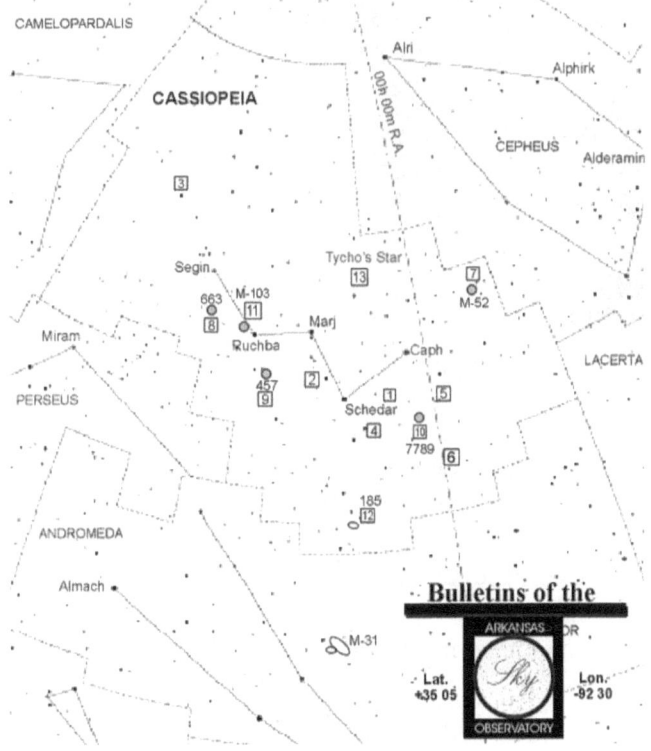

Every deep sky object and every double/multiple star will have a "PERFECT MAGNIFICATION" .....this is the magnification that you should use that will show the object as bright and with as much as detail with possible and still increase its size appreciably so that you can view it comfortably and unmistakably. The rule for determining "optimum magnification" is that: 1) too low power results in sky background glow detracting or diminishing the contrast against the deep sky object; 2) too high magnification darkens BOTH the sky background AND the object; 3) medium magnification can be achieved at which you have MAXIMUM contrast between the object and its darkened background sky. I have found through three decades of direct observing that about 15x per inch aperture, for deep sky observing is PERFECT for most objects. That being said, always remember that DOUBLE or multiple stars require whatever power you can crank out....the seeing conditions are the limiting factor here.

For my complete and comprehensive discussion regarding seeing conditions and sky transparency, see my GUIDE on this subject on the *ASO website*.

With all deep sky objects, avoid attempting to observe when the moon is in the sky, even a very thin crescent, as its brightness in the sky will overshadow the very dim contrast afforded by even the brightest deep sky object; if you see the object at all against moonlight, you will NOT see the subtle outlying areas or the full detail of what is presented.

For detail descriptive lists of the great double stars within Cassiopeia, and as with all of the "GO TO"

TOUR constellation lists, I recommend a good star atlas and/or chart which will list all the finest objects, constellation-by-constellation. One very handy reference guide is the *PETERSON FIELD GUIDE TO THE STARS AND PLANETS*, which features complete lists with declinations, right ascensions, magnitudes, and all pertinent information for you to expand your observing horizons beyond this brief GUIDE. For the many double and multiple stars, refer to the indispensable "*Burnham's Celestial Handbook*", Volume 1 for a complete abbreviated listing.

Truly these extensive Constellation study guides will most definitely put your PC sky program to work for you in the most efficient and enjoyable way possible! As a matter of fact, MANY computerized telescope users are now programming their own "Tours" based on these guides, using each constellation as a separate GO TO Tour for the PC library that can be added in or deleted through the main edit screen on your PC or MAC computer.

We hope you enjoy these comprehensive GUIDES to touring the constellations via your AutoStar or other computer-driven telescope. Each new installment is complete with diagrams, charts and illustrations that you will find nowhere else. Please let us hear YOUR feedback and your observations of each and every constellation after YOU have toured its vast reaches of our skies!

YOUR CASSIOPEIA CONCISE DIRECTORY OF INTERESTING OBJECTS –

Cassiopeia is packed with wonderful galactic - or

231

"open" - clusters that are all members of our Milky Way galaxy. Also be sure and note that there are some outstanding double and multiple stars as well. (For a full discussion on double star observing and their "Position Angles" refer to my brief overview in the "GO TO" tour guide for Lacerta found in *Constellation Guides, Vol. II.* Even though Cassiopeia is lacking in distant galaxies, star clusters and numerous and bright nebulae, there is a remarkable wealth of beautiful viewing set high in the autumn and winter northern hemisphere skies for EVERY telescope, no matter what make nor size. The listing which follows is but a sample of the truly remarkable wealth of wonderful objects held within the boundaries of this not-so-stellar Queen of the sky.

As an added bonus to this installment of them "GO TO" TOUR constellation guides, a complete abbreviated list of the 11 brightest galactic star clusters follows in addition to our regular featured directory of the finest objects for this constellation.

NOTE: from here forward, the reference to "CASSIOPEIA" will be abbreviated (it twists one's fingers to type it too many times!) by the simple designation "**Cass**".

I have chosen the most interesting 14 targets in this CASSIOPEIA "GO TO" TOUR; as with all GUIDES, all objects listed below will be visible in most telescopes (some naked eye) from scopes in size from 3-8 inches; of course larger apertures may "show" an object a bit closer and "better," but frequently a wide field and low power view is more desirable than aperture for FINDING the objects initially. Indeed, I strongly encourage you first FIND the target object, or its approximate location through your GO TO function with your lowest power and then - once IDENTIFIED positively - move up slowly in steps with magnification if necessary. Remember, not all objects "like" magnification. Sometimes better "field of view" (such as the wonderful wide fields provided by snaller telescopes) is desired over light gathering (like an 8-inch) and magnification. Note that your AutoStar or sky program may NOT have every object listed on every constellation GO TO tour....this is intentional. You can access some of the most interesting objects of the sky directly from their coordinates. It is quite simple as you merely enter these coordinates as follows in the 10-step

You will access your FIRST GOTO target - (usually the brightest star in each constellation) - via the command "SETUP / OBJECT / STAR / NAMED....and scroll to "**SCHEDAR**", then press "Enter" and subsequently "GO TO" to move your this bright star. (sky programs will vary)

You may also access the constellation by: SETUP/OBJECT/CONSTELLATION/"Cassiopeia" .....Enter....GO TO, which will subsequently take you to the brighter star Schedar, which makes one of the "bottom" (western-most) angles of the conspicuous "W" shape of this constellation.

OBJECT 1: brighter star - SCHEDAR (alpha Cas) - R.A. 00h 38' / DEC + 56 16 - Mag: 2.2 + 9.1 mag. companion!!
OBJECT 2: wonderful double star! - ACHIRD

(eta Cas) - R.A. 00h 46' / + 57 33 - Mags. 3.6 & 7.5 - ALL SCOPES

OBJECT 3: a great triple - iota Cas - R.A. 02h 25' / DEC + 67 11 - Mags: 4.7, 7 & 8.2 - great triple for 3"!!

OBJECT 4: test for 8" - double lambda Cas - R.A. 00h 29' / DEC + 54 15 - Mags: 5.5 & 5.8 - only 0.5" arc!

OBJECT 5: super double - sigma cas - R.A. 23h 56' / DEC + 55 29 - Mags: 5.5 & 7.5 - A great double star, 3" sep.

OBJECT 6: classic variable - R Cas -R.A. 23h 56' / DEC + 51 07- Mag. 5.4 to 13! Great long period star for all!

OBJECT 7: beautiful cluster! - Messier 52 (ngc7654) - R.A. 23h 22' / DEC + 61 20 - Mag. 7.3, 120 stars!

OBJECT 8: famous variable - "Mira" (Omicron Ceti) - R.A. 02h 17' / DEC -03 12 - Mag: 2.8 to 9.5 in 331 days!

OBJECT 9: nice face-on galaxy - Messier 77 (ngc1156) - R.A. 02h 40' / DEC -00 14 - Mag: 8.9, bright and detailed!

OBJECT 10: galactic cluster - ngc663 - R.A. 23h 55' / DEC + 56 26 - Mag: 10 - over 900 stars!! FAINT ones, though!

OBJECT 11: nice cluster - Messier 103 (ngc581) - R.A. 01h 30' / +60 27 - Mag. 7.4, very small, 60 stars

OBJECT 12: neighbor galaxy- ngc185 - R.A. 00h 36' / DEC + 48 04 - Mag. 9.9 - Distant companion to M-31!

*** BONUS: ***

Your Cassiopeia List of <u>BRIGHT OPEN GALACTIC STAR CLUSTERS</u>!! Here is a list of the richest and brightest star clusters that are visible in nearly ALL telescopes in amateur use! From the smallest richest field refractors to the large permanently mounted observatory telescope, there is a wealth of very rich star clusters in this wonderful area; this listing is appearing in abbreviated form by 1) ngc number, 2) Right Ascension, 3) Declination and, 4) magnitude (visual) only; 5) size in minutes arc ('); and, NUMBER of STARS, followed by a very brief description and in some cases telescope and observing notes:

<u>**457**</u>   01 16 / +58 04 / 7.5 - 10', round 100 stars, very large and pretty in very low power, all scopes!
<u>**559**</u>   01 26 / +63 02/ 7.3 - 7' arc, 60 stars, fairly bright, but stars are a bit dim for smaller scopes
<u>**581**</u>   01 30 / +60 27 / 7.4 - small (5') and rich, M-103, great in all scopes; look for 4 brighter stars
<u>**637**</u>   01 38 / +63 47 / 7.1 - very small, 3' arch, and only 20 stars, very faint - medium power!
<u>**663**</u>   01 43 / +61 01 / 7.1 - 11' arc with 80 stars, all resolvable in 6-inch - nice in all scopes
<u>**IC1805**</u>   02 29 / +61 13 / 7.0 - VERY large open cluster with only 20 stars, 20' arc...almost non-clusterlike
<u>**IC1848**</u>   02 47 / +60 14 / 7.6 - another very spread out (17') cluster with 70 stars, pretty dim stars
<u>**Herschel**</u> 1   03 07 / +63 03 / 7.2 - very large (15') and scattered group, some 30 stars, poor cluster
<u>**7654**</u>   23 22 / +61 20 / 7.3 - SUPER object (M-52), over 120 stars in only a 12' arc cluster

7789   23 55 / +56 26 / 9.9 - VERY faint, cloud-like and large (20'); 900 stars to mag. 11 !!!

7790   23 55 / +60 56 / 7.1 - very small (6') but nice cluster with about 25 faint stars...nice in all scopes

YOUR VISUAL GUIDE TO DEEP SKY OBJECTS IN CASSIOPEIA

Object 1 - Our "Starting" Bright Star –
"SCHEDAR" (alpha Cas) with two tiny companion stars!
At a distance of 175 light years, Schedar offers a lot of star for the money....it is variable (mag. range 2.2 to 2.8, erratic) and it has two very faint companions, one 9th magnitude and the other nearly 14th magnitude, still visible in the 8-inch and larger scopes. Interestingly, these two stars have orbital periods around Schedar among the fastest known. The 9th magnitude star ("B" component) takes only 13 years to orbit the primary star; look for it a whopping 63" arc (larger than Jupiter would appear in the same high power (about 200x) eyepiece) from the brighter star. The faint star will be a challenge, even for an 8" scope and is located only HALF the distance of the easier 9th magnitude star. It orbits Schedar in only 8 years!

Object 2 - Eta Cas (ACHIRD) - Famous and
Achird (pronounced "ACK-eared") is one of the most often observed of all high northern double stars. Its two components are relatively bright (3.5 and 7.2 magnitude) and widely separated (about 12" arc right now, very easy for the 3-inch scopes at medium (40x) magnifications. Look for the fainter star of the two nearly DUE SOUTH right now from

the primary star (position angle 190 degrees). The next few years will exhibit this double star at their WIDEST separation possible in its entire 526-year orbital period as we observe it from Earth, so get out an look at it! LOOK FOR THE DISTINCT COLOR CONTRAST of these two stars, one of the most remarkable in the entire sky! The primary (brighter) star is a brilliant golden yellow, while the fainter star exhibits a distinctive PURPLE color, a wonderful contrast in brightness AND in color for even the smallest telescopes. This star is best viewed with about 15x per inch aperture in any telescope.

Object 3 - Nice Triple Star - Iota Cassiopeia

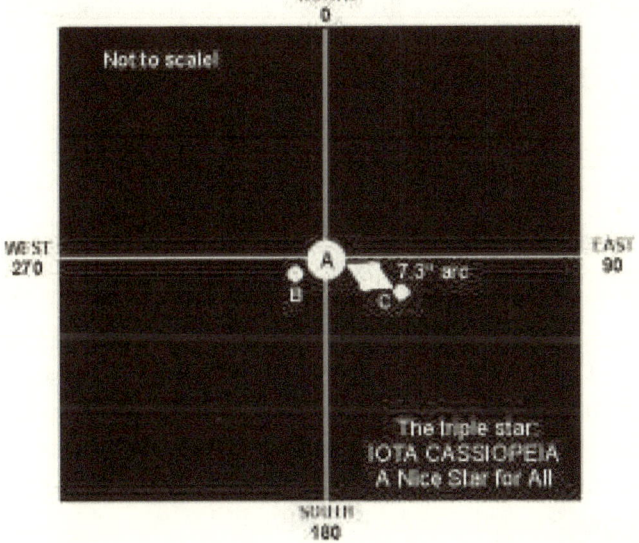

This finder/locator chart should serve as your invitation to check out one of the nicest multiple star systems for small, medium and large amateur telescopes. Iota Cass is a true physical triple system. The primary star is a very yellow-white magnitude

4.7; there are two companions to this star, a clearly blue one ("B") only 2.3" arc away (easily resolvable at high magnification - about 25x per inch)a t magnitude 7.0 at position angle 240 degrees (nearly due west) and a very RED fainter companion ("C", magnitude 8.2) some 7" arc nearly opposite the brighter star from "B". All three stars are bright enough to see clearly in small telescopes, but a steady night at least a good 3-inch is required to split the "A" and "B" components; the "C" star should be relatively easy in all small telescopes,, although "B" may not be seen at any power. Give it a try!

Note that "B" requires about 840 years to orbit the primary as seen from Earth; "C"'s period is still undetermined, but best estimates give an orbital period of greater than 4,000 years!

Object 4 - A Test Object for the 8" and Larger Telescopes - Lambda Cassiopeia
Here is a toughie for 8-inch and larger telescopes. This double star will require pretty high magnification (about 320x) and very steady skies when Lambda is nearly overhead, or highest in the sky. This double is a VERY blue hot pair of equally matched stars, magnitudes 5.5 and 5.8 visually. They are separated ONLY by 0.5" arc and quite a test for larger amateur telescopes. Because of the brightness of this pair, they should be easy to locate and recognize at low power; then begin moving up in magnification to over 300x (or whatever it takes!) to show some dark sky between them. I have split them with difficulty with an 8-inch, (well collimated and a very steady, foggy night!) and pretty easily repeatedly with a 12" Schmidt-

Cassegrain. My 6" Unitron showed these stars as elongated, but not cleanly separated at any power. Look for the stars in an almost NORTH-SOUTH orientation from one-another.

Object 5 - Great Double Star - Sigma Cas
This is a pair of stars that deserve observing with a 5 inch and larger scopes. The primary is magnitude 5.5 while the fainter companion is 7.5 in Position Angle 326 degrees (almost in a north-south orientation). These very blue-white young stars are separated by 3.0" arc, so they are clearly resolvable in the 3" scopes with medium (about 150x) magnification.

Object 6 - A Classic Long Period Variable Star - R Cassiopeia
Here is a very nice variable stars, one of the finest long-period variables (LPV) in the sky. R Cas will go from a bright naked eye 5.4 magnitude (easily observable when at its brightest phases in all telescopes!) to a dim magnitude 13, still attainable in the 6-inch scope and even with a 3.5" scope for the keen eyed! The light changes from brightest to brightest again require a total period of 431 days, so this is a long-term (more than one year!) project if you choose to follow this star's changes and chart your own light curve. The curve shown below is from estimates of observers in the American Association of Variable Star Observers (AAVSO - www.aavso.org) over many years of observing.

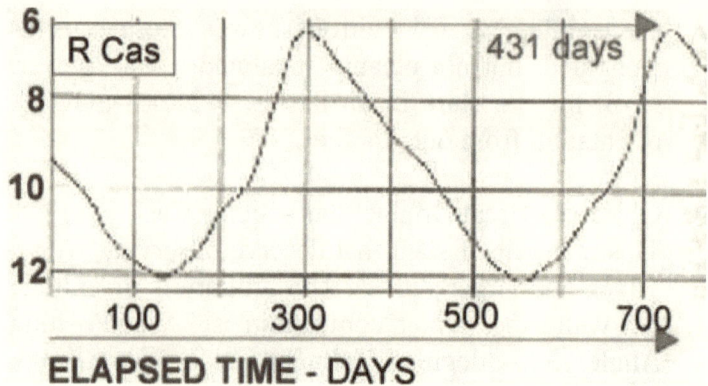

ELAPSED TIME - DAYS

Since the light changes so slowly as this huge star pulses out (from thermal pressure) and inward (from gravity), that observations should not be made but every two weeks or so....over-observing usually leads to spurious brightening or changes that you record that "really" don't exist! When the star is at its brightest, you want to use the "locator" chart from the AAVSO, chart "A", available for free downloading at http://www.aavso.org/vsp (enter "r cas" for the name).

As it dims below 9th magnitude, be sure an move up in your chart to the narrow field, high power "C" chart which gives accurate comparison magnitudes for very faint nearby stars with which to compare. This chart is also available at http://www.aavso.org/vsp (enter "r cas" for the name).

Merely right click on the chart and save it to file on your computer; call up the file and resize it to fit your standard 8.5 x 11 inch paper size and re-save in that new format. Then simply print out each of these charts, laminate for outdoor use and observe away! (see my complete discussion on variable stars in GUIDES Observing/General on the Arkansas Sky Observatories website: www.arksky.org.

Object 7 - A Jewel Box of the Sky...Messier 52....Splendid Open Galactic Cluster
Just looking at the beautiful Lowell Observatory photography of Messier 52 below reminds me of sparkling diamonds scattered onto a black velvet mat....some larger and more brilliant than others, but all necessary to make this splendid sight what it is.

MESSIER 52 - A Galactic Cluster

Messier 52 is located in the extremely star-rich arms of the winter/fall Milky Way, home of many such clusters nearby. The British Admiral Smyth recorded his impressions of the cluster as an irregular patch of stars that reminded him of a bird in flight, wings outstretched! Near one of the "apex" points of this triangular deep sky object is a wonderful orange-red star that is clearly a contrasting color to the remaining 120 stars, all in the 10th magnitude range. Note that in the western part of the cluster are two more brighter stars as well as one single brighter star on the opposite (eastern) side. This is a very compact but fairly large (12' arc) cluster that can be seen even in binoculars or your finder as a fuzzy patch (magnitude 7.3). By all means observe this cluster ONLY with a wide field and low magnification for your best views.

Even as I write this I have just come in from an early morning "test" of various telescopes on this cluster. It is an interesting object. With the 24" scope, the cluster is not at all impressive and looses its true cluster form in the narrow field of view. A 10" Schmidt-Cass shows the entire cluster very well at low power, but it appears large and scattered, much as it does in the 8" telescope; a wider field, low power scope shows it perhaps better than all others, with all stars being visible that are in other larger scope, but with a brightness and image scale that is much more appealing to the eye! Thus, if you are using a 3" to 6" telescope, this cluster will become one of your all-time favorites!

Object 8 - Another Beautiful Galactic Cluster – NGC 663 (and nearby ngc654)
Here is a double treat! NGC 663, at magnitude 7.1 is a fairly large (11' arc) round conglomeration of about 60 stars; some of the group are fairly bright

(8-9 magnitude), and the entire cluster is a beautiful sight in low power views of all telescopes; it is particularly pleasing and rich with very faint stars in the 5" and 8" scopes. Just to its north is a much less populated cluster, and about half the size, ngc654, which has some 50 stars of about the 11th magnitude. This latter cluster is a nice object - and challenging - for a small APO scope and is quite a nice sight in the 5". If using a very wide field eyepiece, you can get both these objects in the same field of view (this makes a beautiful grouping in the smaller scopes!), with ngc654 about 1/2 degree NORTH and slightly to the east of ngc663. This pair is well shown in the wide-field photograph courtesy of Lowell Observatory's 13" sky camera.

Galactic Clusters
NGC 663 (bottom) NGC 654 (top)
Object 9 - Another Galactic Star Cluster - NGC 457

This is a fantastic object often overlooked by amateurs, located about 4 degrees from "the middle of the 'W'" asterism of Cassiopeia. NGC 457 contains over 100 stars, all of which are discernible to the 4-inch and larger scopes. Even the 3" can make out many of the beautifully faint individual stars as the "twinkle" in and out of visibility, just at the threshold of your limiting magnitude.

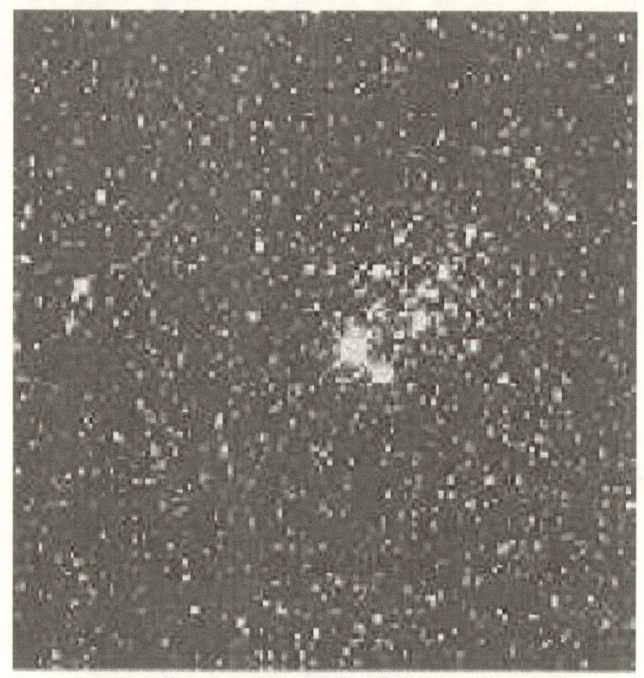

A Jewel Box in Cassiopeia
NGC 457

As seen in my photograph, there is a brighter star, magnitude 7.2, just on the southern edge of this cluster. This yellowish star is actually a member of the cluster and NOT a foreground object as seen in so many clusters containing seemingly brighter members. With a total integrated magnitude of 7.4 visually, this cluster is visible as a bright "smudge" in the wide field APO scopes and with partial resolution in the 90. It is a wonderful sight in the 8" scope, fairly large and uniformly filled with very faint stars except for the two bright stars in its southeastern quadrant.

<u>Object 10</u> - Galactic Cluster NGC 7789
Here is a beautiful object for the 3" and larger scopes; even a good small APO will clearly show this cluster as a very faint patch of "star dust" between the bright stars Rho and Sigma Cas. As you increase aperture, the more stars that are visible....indeed, up to 1,000 stars exist within the gravitational bounds of this remarkable and distant (6,000 light years!) object. This is a very rare and unusual galactic cluster. has recently been found that this is one of the OLDEST ones known.....nearly 2 billion years old....older than most, but not nearly as old as the "globular star clusters" (see my article on globulars http://www.weasner.com/etx/ref_guides/globulars.html). Thus there is some discussion as to whether this might be a very "young" and sparse globular, or a very old open cluster.... NGC 7789 is about the apparent size (30' arc) as the full moon, so it requires a wide field of view and VERY dark skies to appreciate its splendor! (photo below courtesy Lowell Observatory).

The Extremely Rich 900-star
NGC 7789

OBJECT 11 - A "non-Messier (?)" Messier Object!
- Messier 103

There is much disagreement as to whether this very
small (5' arc) and sparse (about 60 stars) was
actually included by Charles Messier in his original
list of "faint fuzzies" that he compiled to prevent
confusion in his search for new comets. Most agree
that Messier recorded (not "found"....most, if not
all, of the "M-objects" were known long before
Messier ever began his illustrious career as a comet
hound) a total of 104 objects, but two are uncertain
as to whether they were actually the RIGHT objects
on his list. Messier 103 (ngc581) seems to be the
actual FINAL M-object, while Messier 104 (the
Sombrero galaxy) has been added in more recent
times. Messier 103 is in the same very wide field of
view as Delta Cas (RUCHBAH), some one degree

southwest away. There are four brighter stars (at least 9th magnitude) that should be visible in the 3" scope against the fuzzy faint glow of the remaining four dozen or so stars. BE SURE and look for the scarlet-colored very red (9th magnitude) and blue (mag. 6.2) double star (Struve 131) embedded directly in this nice cluster! The two stars are a wonderful color contrast and are separated by a full 14" arc, so it should be an easy object, both in terms of separation and in magnitude for all scopes at medium-high (about 25x per inch) magnification.

OBJECT 12 - Elliptical Galaxies - NGC 185 / 147- Companions to Famous Andromeda Galaxy
Like ngc105 and Messier 32, ngc185 & 147 are gravitational partners to the famous Andromeda Galaxy, Messier 31 (see: the Constellation Guides Andromeda). When you compare this map to the proximity of Andromeda, you will see how that is indeed possible; all these as well as the Milky Way and a few others comprise what is known to astronomers as the "Local Group", a gravitationally-bound cluster of galaxies in this "neck of the Universe."

Look for ngc185 as a VERY faint (magnitude 11.0), 2.2' arc round glow in a very low, wide-field eyepiece, just barely discernable from an out-of-focus star; this can be seen barely with a four-inch and with difficulty in the 5" scope. Off to the north, and a little west of this one (about 50' arc total) you will see an even fainter galaxy (147) at magnitude 12, just at the threshold of the 5" scope....the 8" and larger are really needed, but to get them both in the field of view will require the 32mm 2" Super Wide Angle or similar. Do not expect to see ANY detail in even large amateur telescopes, and resolution is only accomplished photographically with scopes over 100" diameter!

WANDERING ABOUT....YOUR NEW "USER OBJECT" IN CASSIOPEIA

You are about to key in an object into your USER OBJECT LIBRARY that you: 1) did not see; 2) will not see now; and, 3) likely NEVER will see....but it IS worth putting in your library nonetheless. Refer to the chart below (north at top, east at the right, just as with your catadioptic telescope with the diagonal mirror/prism in place) to locate what you cannot see:

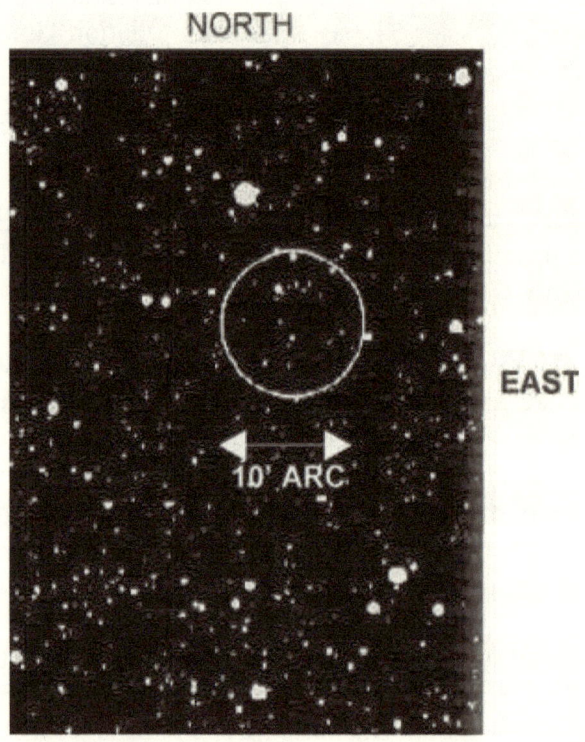

In the year 1572, a star exploded in the sky - a "supernova" - that filled our Milky Way galaxy with light; from Earth, it was brighter than even the brilliant planet Venus, and nearly that of the quarter moon. It is one of only four know (since recorded history, at least) supernovae to have occurred in our galaxy. Although not first "discovered" by **Tycho Brahe**, he was the first to measure and carefully record its brightness and coordinates on <u>November 11 in 1572</u>. He noted that for over two weeks, the "new star" had been seen in mid-day, and remained visible to the naked eye for over 16 months! But....no matter how bright this star WAS, there is no trace of it today. Only scant traces of a nebulous cloud can be seen with the world's giant telescopes....no star is left brighter than around 20th magnitude.

Nonetheless, you can point out this special place in the sky where the GREATEST EXPLOSION in all of recorded history was observed over 400 years ago. The coordinates of "TYCHO'S STAR" are: Right Ascension: **00h 22.0m** ; Declination: **+ 63 degrees 52** minutes. The center of the white circle indicates the position accurately. This is not only a part of the sky that you should observe in a rather "humbling way," but also one that is a terrific "talking point" at any star party or club outing!

On your sky program, go to: "Select/Object [enter]...." scroll down to "User Object" [enter]. Now enter the coordinates given above for "Tychos", key in the coordinates on your PC program or, using the number keys on AutoStar. After entering the coordinates and pressing "Enter" yet again, scroll down one and you can list the magnitude of the object as "20"[Enter].

*   *   *

Stay tuned.....It's a ZOO out there! We have whales, dolphins, fish and even flying horses....so what's a little GIRAFFE going to hurt? That's right: in coming installments of our computerized telescope "GO TO" guides to the *Constellations* we will take a look inside (figuratively, thank goodness!) great whales, sea goats, the night's sky's long-necked spotted furry friend. Why a giraffe in the sky, you ask? Me too....

Nonetheless, we will explore this often-overlooked constellations without nary a named star nor Messier object, but having a few worthy galaxies and some very nice and challenging double/multiple stars! In addition, this constellation offers a couple of "all night observable" variable stars since it - just like Cassiopeia - is circumpolar from northern latitudes, meaning that it encircles the north celestial pole never rising nor setting throughout the 24-hour day.

Good Observing and may the stars serve as your sentries as you explore the frontiers of space!

*   *   *

Chapter Ten

CETUS

....look! in the celestial sea! It's a mermaid! No, it's a whale! Well, maybe it's really a SEA MONSTER!

This *Constellation* guide, "GO TO CETUS" of the series *Constellations* for telescopes, gives us the curious constellation of Cetus, located just below the celestial *ECLIPTIC*, so the planets, moon, and sun are sometimes nearby, but never within the boundaries of this fairly southern constellation. It is a largely ignored constellation as are many in this area. It IS more sprinkled with brighter stars than other nearby neighbors such as Aquarius, and it contains no less than 27 galaxies bright enough for common telescopes! Indeed, all 27 of these are within the grasp of an 8-inch and larger telescopes; some 20 within reach of the 6-inch, about 14 with a good 3-inch and at least five with smaller scopes!

In addition, Cetus holds a WEALTH of multiple stars, the second-most famous variable star, *MIRA* (second only to *ALGOL* in Perseus), a fantastic planetary nebula, many star groups and other celestial sights. So how many hours have YOU spent in the large belly of the whale?

Or is it really a "whale?"

In Ethiopian lore, the princess *Andromeda* was the daughter of *Cassiopeia* and *Cepheus* (see my Constellation Guide Cassiopeia here on this ASO website). Unfortunately her mother was so vain that she thought herself more beautiful than the daughters of *Nereus*, one of the many gods of the

251

sea. So Cassiopeia was hit by the anger and the revenge of the god *Poseidon.* To punish the mother, Andromeda was chained to a rock of the coast as a sacrifice for a sea monster, now enshrined as the constellation CETUS (more aptly now configured as a mere "whale!"). Yet she was saved from certain "sea monster death" by the hero, Perseus. After rescuing her he demanded Andromeda as his wife (which the parents gladly accepted, although it was rumored that Perseus was also perhaps "involved" with the wife, Cassiopeia.

It seems that in later lore this "sea monster" has been relegated to a mere Whale of a monster!

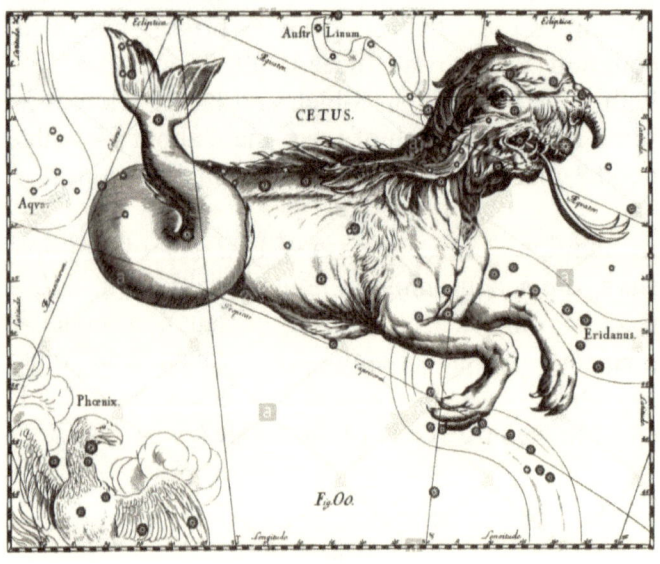

Cetus the Whale or Monster Whatever it Might Be
By Johannes Hevilius, 1687

This region is a very minor area of the sky in terms of bright showcase objects, Messier objects (there is only one) and bright stars. However, the constellation makes up in SIZE what it lacks in remarkable objects. So....for the patient observer, spending a little time "going to...." objects in the whale will be VERY rewarding, indeed!

For that reason, I have included a very brief listing of the entire group of 27 galaxies within Cetus that are observable with at least some of our telescopes. There is also an increased number of fine double and multiple stars on our object list for this constellation as well. As a matter of fact, there are so MANY good multiple star here that I highly recommend those interested in the beauty and challenge of these objects to consult with the listing at the beginning of the Cetus chapter in Robert Burnham Jr.'s *Celestial Handbook*, Volume One. There you will find a concise directory of stars, magnitudes, position angles of major components, separations of the stars and other brief abstracted information.

Use the finder chart following for locating many of the GO TO objects in the constellation of Cetus; if using a computer planetarium program, you are encouraged to plot the objects on your screen for higher resolution than this chart provides.

CONSTELLATION GUIDES:
"Go To" CETUS the "Whale"
Or Perhaps a Monster of the Sea!

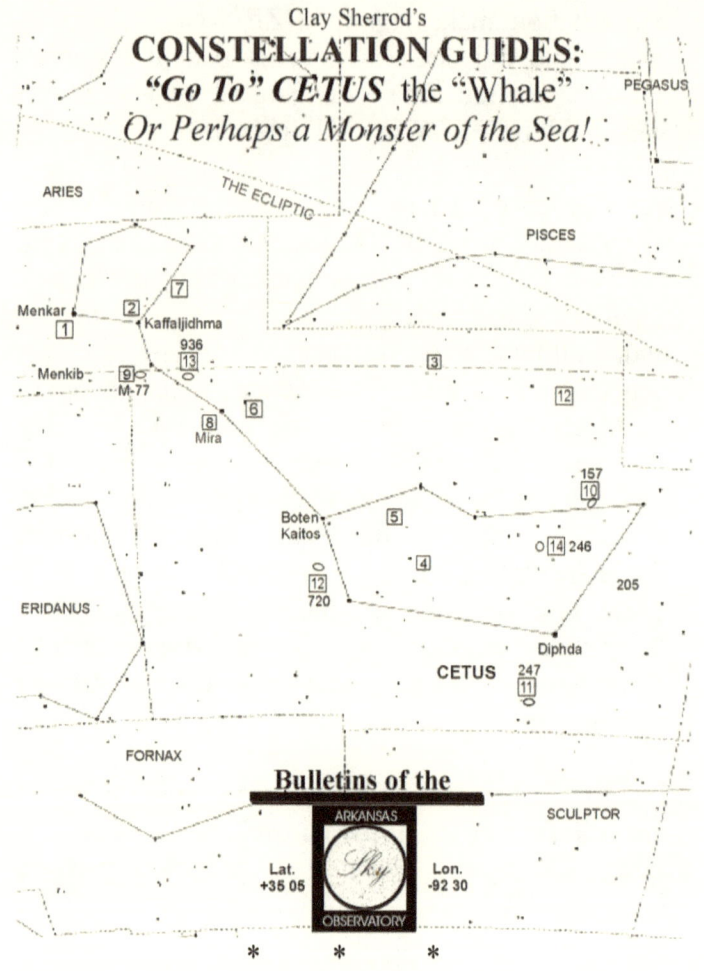

Cetus (pronounced "SEE-tus") is rather uninteresting naked eye constellation of late summer and early fall skies. To most observers who have not spent much time exploring the Whale, it is conceived, incorrectly, that Cetus is located much farther south in the northern hemisphere skies than it actually is. Its highest extent reaches into +10 degrees declination, while the lowest point only sinks to -25 degrees south. Thus, nearly all objects are visible within this large constellation (its breadth extends form 00h to over 03 hours in right ascension - more than 45 degrees of sky east to west!

Cetus rises just south of due east for mid-northern latitudes of about 35 degrees north, its first (westernmost) star Diphda rising about 9 p.m. local time on about September 1. Low to the south, it reaches the meridian, or highest point in the sky, culminating at 2:30 a.m. on the following morning. Midnight culmination - when the center of the large constellation is on the meridian at midnight each year - occurs always on about October 15.

It appears that the great whale is plagued a bit by a pesky "water snake," Eridanus to its south, as the long serpent writhes northward and actually "makes contact" through a mutually share bright star - Pi Ceti - in both drawn outlines of the Whale and the Water Snake!

As with every "GO TO" tour guide, each GO TO object in Cetus is discussed for your telescope regarding the type of conditions necessary for you to view it optimally for discern the very faintest details.........magnifications and aperture necessary for most objects, and much, much more. This is YOUR complete guide to get you on your way to exploring the best (and few!) objects in this small constellation. The following listing of "BEST" objects contains the finest or most interesting from my own observing experience and preference.

Use the attached star chart (above) and the following Guide as an excellent reference for your next star party itinerary, or a beginning for further study into the thousands of objects visible in this part of the sky. The chart gives the outline of the major constellation as well as the approximate locations of all objects discussed in this guide.

Truly these extensive *Constellation* study huides will most definitely put your PC sky program to work for you in the most efficient and enjoyable way possible! As a matter of fact, MANY AutoStar and other sky program users are now programming their own "Tours" based on these guides, using each constellation as a separate GO TO Tour for the planetarium program and sky library that can be added in or deleted through the main edit screen on your PC or MAC computer.

We hope you enjoy these comprehensive GUIDES to touring the constellations via your AutoStar or computer-driven telescope. Each new installment is complete with diagrams, charts and illustrations that you will find nowhere else. Please let us hear YOUR feedback and your observations of each and every constellation after YOU have toured its vast reaches of our skies!

YOUR CETUS CONCISE DIRECTORY OF INTERESTING OBJECTS –

There is a plethora of often-overlooked and very nice deep sky objects in Cetus. In addition, there are some outstanding double and multiple stars as well. For those who think that this part of the sky is void of interesting objects, spend some time here and you will change your mind. The listing which follows is but a sample of the truly remarkable wealth of wonderful objects held within the boundaries of the great Sea Monster. If you enjoy beautiful and challenging double/multiple star observing, there are some wonderful targets for you in Cetus. For a full discussion on double star observing and their "Position Angles" refer to my brief overview in the "GO TO" TOUR guide for Lacerta located in the GUIDES/Constellation Guides here on the ASO website .

I have chosen the most interesting 14 targets in this CETUS "GO TO" tour; as with all guides, all objects listed below will be visible in most telescopes (some naked eye) from scopes in size from 3-8 inches; of course larger apertures may "show" an object a bit closer and "better," but frequently a wide field and low power view is more desirable than aperture for FINDING the objects initially. Indeed, I strongly encourage you first FIND the target object, or its approximate location through your GO TO function with your lowest power and then - once IDENTIFIED positively - move up slowly in steps with magnification if necessary. Remember, not all objects "like" magnification. Sometimes better "field of view" (such as the wonderful wide fields provided by smaller telescopes) is desired over light gathering (like an 8-inch) and magnification. Note that your sky program may NOT have every object listed on every constellation GO TO tour....this is intentional. You can access some of the most interesting objects of the sky directly from their coordinates. It is quite simple as you merely enter these coordinates as follows in the 10-step process described in **Using a Computerized Telescope**, at the start of Vol. I

You may also access the constellation by: SETUP/OBJECT/CONSTELLATION/"Cetus"..... Enter....GO TO, which will subsequently take you to the brighter star Menkar. (sky programs vary)

Following is the concise object list for your "GO TO" TOUR of CETUS; you may wish to find many of the objects from the AutoStar or PC sky Library (for example, you can easily go a fairly bright galaxy, Messier 77, if you pull up "Object/Deep Sky/Messier Object/..then type in '77'...." and then press "Enter", followed by "GO TO" to access this nice spiral galaxy. On the other hand, if you want to experiment and become a "better computer user" try entering the exact R.A. and DEC coordinates of that object as given below after holding down the MODE key. You will find the accuracy of entered GO TO's to be somewhat less than those stored in your sky program, but the capability of acquiring unlisted objects is fantastic!

You will access your FIRST GOTO target - (usually the brightest star in each constellation) - via the command "SETUP / OBJECT / STAR / NAMED....and scroll to "MENKAR", then press "Enter" and subsequently "GO TO" to move your this bright star. (your sky program may vary)

You may also access the constellation by: SETUP/OBJECT/CONSTELLATION/"Cetus"..... Enter....GO TO, which will subsequently take you to the brighter star Menkar, near the northeastern edge of this constellation.

OBJECT 1: brighter star - MENKAR (alpha Ceti)

- R.A. 02h 59' / DEC + 03 54 - Magnitude: 2.5 + bonus double star!

OBJECT 2: bright double star - KAFFALJIDHMA (gamma Ceti) - R.A. 02h 41' / DEC + 03 02 - Mags: 3.6 & 6.2 nice!

OBJECT 3: "double" triple - 42 Ceti - R.A. 01h 17' / DEC -00 46 - Mags: 6.4 & 7.3 - 3rd star is too close to see!

OBJECT 4: nice double! - h2036 - R.A. 01h 18' / DEC -16 04 - Mags: 7 & 7! Nice yellow evenly matched pair!

OBJECT 5: great 4-inch test - B399 - R.A. 01h 23' / DEC -11 10 - Mags: 6.5 & 10 - A great double star, 1.6" sep.

OBJECT 6: nice easy double! - 66 Ceti -R.A. 02h 10' / DEC -02 38 - Mags: 5.7 & 7.7 - wide spacing, great for all!

OBJECT 7: beautiful double - Nu Ceti - R.A. 02h 33' / DEC + 05 23 - Mags 5 & 9.5 - A real treat!

OBJECT 8: famous variable - "Mira" (Omicron Ceti) - R.A. 02h 17' / DEC -03 12 - Mag: 2.8 to 9.5 in 331 days!

OBJECT 9: nice face-on galaxy - Messier 77 (ngc1156) - R.A. 02h 40' / DEC -00 14 - Mag: 8.9, bright and detailed!

OBJECT 10: very small galaxy - ngc157 - R.A. 00h 32' / DEC -08 40 - Mag: 11, "do-able" with the 3-inch scope and up!

OBJECT 11: huge galaxy, but faint - ngc247 - R.A. 00h 45' / DEC -21 01 - Mag. 9.9 - Very large and spread out.

OBJECT 12: elliptical galaxy - ngc720 - R.A. 01h 51' / DEC -13 59 - Mag. 10.5 - very small, challenge for 5" scope

OBJECTS 13: barred spiral and faint spiral - ngc936 & 941 - R.A. 02h 25' / DEC -01 22 - Mags 10.5 & 12.5 - NEAT!

OBJECT 14: planetary nebula - ngc246 - R.A. 00h 45' / DEC -12 09 - Mag. 8.5, central star is 11.3! Nice object!

*** BONUS: ***

Your Cetus List of OBSERVABLE GALAXIES!! Here is a list of galaxies that might be visible in at least an 8-inch scope, and most are visible with some difficulty in a 5-6-inch as well; this listing is appearing in abbreviated form by 1) **ngc** number, 2) Right Ascension, 3) Declination and, 4) magnitude (visual) only, then followed by a very brief description and in some cases telescope and observing notes:

151 00 32 / -09 58 / 12.2 - elongated spiral, a target for a 6-inch and larger scopes only

157 00 32 / -08 40 / 11.0 - nice face-on spiral, can be viewed in 3" and larger apertures; brighter center

210 00 38 / -14 09 / 11.6 - larger spiral, somewhat elongated; viewable in 5" but unlikely in 3" scope

247 00 45 / -21 01 / 9.5 - very large spiral (see description below) but very spread out and dim

255 00 45 / -11 45 / 12.4 - small round, very faint star-like spiral, appears nearly stellar in 5"

309 00 54 / -10 13 / 12.5 - VERY faint spiral, visible in LX scope only

337 00 57 / -07 51 / 12.1 - tiny distant spiral, mistaken for a star even in 8" scope

428 01 10 / +00 43 / 11.6 - peculiar spiral, can be spotted in 5", fairly large blur and dim

578 01 28 / -22 56 / 11.2 - oval spiral galaxy, can be barely seen in a 3-inch

584 01 29 / -07 07 / 11.2 - VERY tiny elliptical, can easily be mistaken for a star; requires 5" and up

596 01 30 / -07 17 / 11.5 - fairly bright elliptical, good target for 3", very small and star-like

636 01 37 / -07 45 / 12.4 - EXTREMELY tiny elliptical, visible only in 8" scope; starlike and no detail

720 01 51 / -13 59 / 10.5 - another elliptical, but bright enough for a 3-inch and above, small

779 01 57 / -06 12 / 11.3 - very thin sliver of an edge-on spiral galaxy; glimpse in 3", nice in 8-inch

788 01 59 / -07 03 / 12.5 - very small and faint spiral, possible in 5"

864 02 13 / +05 45 / 11.4 - larger spiral can be glimpsed in 3" but fairly nice in 8" scope and larger

895 02 19 / -05 45 / 12.2 - small oval spiral, glimpsed in 5" scope, not much better in larger scope

908 02 21 / -21 27 / 11.1 - fairly large and brighter spiral, easy target in 3" scope and large in 8"

936 02 25 / -01 22 / 10.5 - nice brighter barred spiral, seen in 3" but nice detail in 8" and larger scopes

1022 02 36 / -06 53 / 11.2 - very small but brighter roundish spiral; glimpse in 3-4 inch scope

1048 02 38 / -08 45 / 12.4 - larger, but VERY faint, a target probably only for the 8-inch+ telescope

1052 02 39 / -08 28 / 11.2 - tiny elliptical but viewable in a 3-inch; not much different in larger scopes

1055 02 39 / +00 16 / 11.4 - larger, very elongated spiral, nice in 8" scope

1068   02 40 / -00 14 / 8.9 - Messier 77 (see description below)
1073   02 41 / +01 10 / 11.3 - nice round barred spiral; see detail on dark night in 5" and larger scope
1087   02 44 / -00 42 / 11.7 - elongated faint spiral, visible in 5-inch and larger
1090   02 44 / -00 27 / 12.5 - very faint, visible only in 8-inch+ telescope with difficulty, star-like

YOUR VISUAL GUIDE TO DEEP SKY OBJECTS IN CETUS

Object 1 - Our "Starting" Bright Star - "MENKAR" (alpha Ceti) **BONUS nearby DOUBLE "triple" star!**
There is a curious arrangement of many 2nd to 3rd magnitude stars spaced evenly throughout our skies that are all in the range of 75 to 175 light years distance, relatively close by cosmic standards. Alpha Ceti, or "Menkar" as it is called is one of them at 150 light years. Menkar in itself is a relatively mundane object, but only 16' due NORTH of this brighter RED star is the 5th magnitude star "93 Ceti" which makes an optical double (not a true "physical" double in that it is bound gravitationally) with Menkar. The star is clearly blue in medium powers (about 70x) and makes a wonderful sight with Alpha Ceti! The blue-red contrast is just remarkable in all scopes, from 3 inch to much larger scopes.

Now....for 3-inch and larger telescope users, here is a challenge, one that should be easily met with a 6-inch. Look midway between Menkar and 93 Ceti....and just a BIT to the east. Here you will find two very faint (11th magnitude) stars that are separated by about 1.7' arc, pretty wide and easy for all scopes. However, look very closely at the 11th magnitude star that is SOUTHERNMOST and you will see that it, too, is double and much closer! It will appear as two very faint stars, both about the same magnitude; use medium powers (about 50x per inch) to best view these three stars.

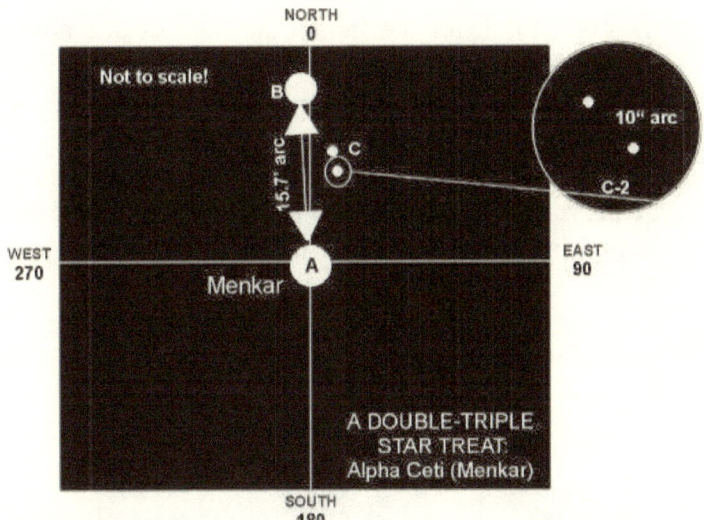

Object 2 - Gamma Ceti - Close and Beautiful Double Star

Here is another one of those "neighborhood stars", some 70 light years away. Gamma Ceti is a gorgeous double star that is easily resolvable in a 3-inch at high magnification, being separated by about 2.7" arc. The brighter star, brilliant white-yellow, is magnitude 3.6 while the fainter - about northwest (Position Angle 293 degrees) - is 6.2 in brightness. See if you agree that earlier observers noted the brighter star as "yellow" and the fainter a "brownish-blue."

Object 3 - Nice Triple (but you won't see all three!)

Star - 42 Ceti

Here is one that provides a good view while allowing your imagination to work a bit for you as well. 42 Ceti is a triple star, two of which are visible in a 3-inch and larger telescopes. These two stars - magnitudes 6.4 & 7.3 are a tough test (1.6" arc separation) for the 3" scope with about 120x, but should split cleanly. The brighter star should appear very red compared to the more yellowish fainter member. Look for the fainter star almost due north (Position Angle 9 degrees) of the primary star. The fainter star is ALSO a double star, both magnitude 7.5, but is so close (only 0.1" arc) that only the LX 200 16" has a ghost of a chance to show elongation of these two, but even it will not cleanly split the star. Thus, for most of us, we must use our imaginations when viewing this fainter star; if our eyes were big enough they would see a slightly fainter object nearly DUE WEST of the image that you see!

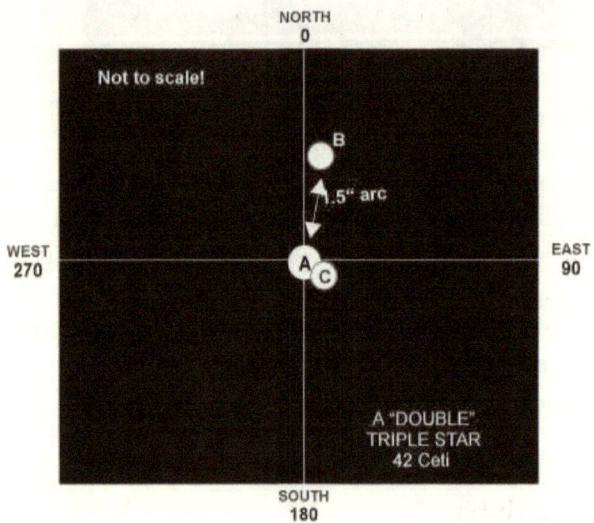

264

Object 4 - A Very Nice Double Star, and a Great Object for a good 3-4 inch telescope - "Herschel 2036"

This star is bright enough to easily be seen in a 3-inch, but the two components are two close to one-another to split it; under very high power (about 125x) with a 2.5" scope, the star will show elongation however. The two stars are an identical match, both red giants and 7th magnitude. They are arranged in a nearly north-south orientation from one-another, so are easy to sight in on.

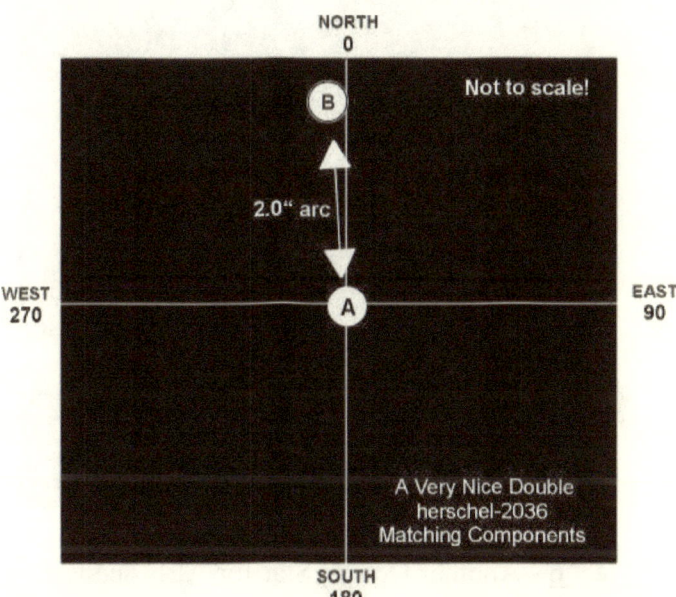

Object 5 - Great Double Star - "Barnard 399" - Not easy, but a great sight in larger scopes

This is a pair of stars of much contrasting brightness. The primary is magnitude 6.4 while the fainter companion is 10.3 in Position Angle 302 degrees (about northwest of the brighter star). The separation further challenges users of smaller telescopes, but a good 3-inch "should" be able to make out this faint companion under about 120x. It is quite easy in 5-inch and larger scopes and makes for a wonderful sight if you can get the magnification up to around 200x on a very dark night. With this kind of power, you should begin to make out the scarlet red color of both stars, as they are very late spectral type "K" stars. Use the finder chart I have prepared below to help in locating the faint and close companion.

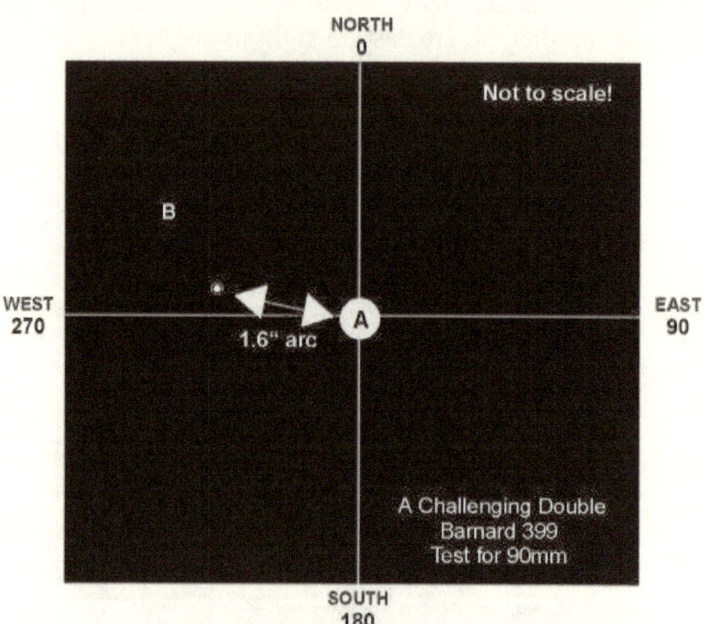

Object 6 - Another Double Star for ALL Scopes! 66 Ceti

Here is a GREAT and easy double star for small telescopes but also well worthy of observing at low magnifications with the larger scopes as well. The very yellow pair 66 Ceti has a primary star magnitude 6.1 and a companion star about magnitude 7.5 a wide 16" arc southwest.

Object 7 - Another Super Double for ALL....Nu Ceti

No matter what size scope, this is a very nice double star for all. Nu Ceti is quite bright at magnitude 5.2 with a companion star magnitude 9.4 about 8" arc away. Although this will be fairly easy with medium low (about 15x per inch aperture) in the 6-inch and larger scopes, it is a challenge for 2-3 inch telescopes due to the faintness of the secondary star, located almost due west of the brighter star. Once Nu is found, up the magnification to about 30x per inch to increase the contrast between the sky and this bright star; the proximity of 5th magnitude Nu will hamper easy acquisition with smaller scopes, but perseverance will prevail!

Object 8 - The Second-most Famous Variable Star of the Sky - Omicron Ceti - "**MIRA**"

No doubt the most popular of all variable stars is "the Demon Star," *ALGOL* in the constellation of Perseus in the northwestern sky. See at the same time yearly is another "possessed" star in Cetus, which just so happens to be the second-most famous variable star of the sky! Omicron Ceti, or "MIRA" is also known as "*the wonderful*" and "the *Terrible*," depending on which school of philosophy your ancestors belonged to. It is the brightest and most notable of all long-period pulsating red giant stars, or the LPV's (see my complete discussion on variable stars in the ASO Guides: *Observing Variable Stars*). In addition, Mira is one of the most "predictable" of all variable stars, although you will clearly see from the light curve of 3.5 years below that the following attributes DO change from time to time: 1) MAXIMUM - you can see that it sometimes barely exceeds 4th magnitude while at others approaches and even reaches 2nd magnitude; 2) PERIOD - although its "average" span from brightest through dim and back again to brightest (its "cycle") is 331 years, note that it does vary substantially from cycle to cycle; 3) LENGTH OF MINIMUM - you can see that sometimes the "valley" at minimum appears to be cut dramatically short, while at others it seems to drag on for a while....ALSO note that the minimum brightness is nearly always consistent - about magnitude 8.8 average.

Regarding maximum brightness, it is highly unpredictable. In 1779 the star rose to 1st magnitude. Thus, through most of its light cycle, the star can be monitored with the naked eye; with standard 7 x 50 or 10 x 50 binoculars, it can be observed throughout the 331 day cycle. Do not attempt to make estimates (which the AAVSO welcomes, by the way, by send through e-mail to: www.aavso.org) more often than every two weeks for best accuracy. "Over-observing" tends to introduce spurious light changes that are really not there!

Mira is an ideal star for smaller telescopes (as well as all other telescopes and binoculars). However, because it is so bright, a very wide field is necessary to obtain comparison stars within the same field of view while estimating brightness. For this, the American Association of Variable Star Observers (AAVSO)provides the naked-eye/ finderscope "a" chart: https://www.aavso.org/apps/vsp/ . Note for these charts, simply type in the NAME of the variable at top to generate your choice of chart – when the star approaches minimum it is best to "zoom in" on the field with the high magnification "g" chart to access stars in the 8th and 9th magnitude range. These charts are available by merely clicking either of the aforementioned URLs, saving the document to file, then resizing and save again. Print out to fit your paper and you have a precise and field tested comparison and finder chart for this star.

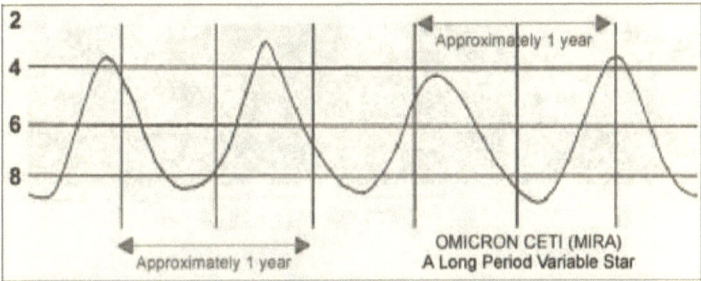

OMICRON CETI (MIRA)
A Long Period Variable Star

The star Omicron Ceti has so impressed observers throughout the history of mankind that it has been coined "wonderful" and "terrible" by all who have noted its variability. It is indeed, both. Mira - from which all other similar red giant pulsating variables have so been designated as "Mira-type" - is a HUGE red giant that pulsates from internal pressure countered by its enormous gravity....it is this pulsation that results in the visible light changes that we see here from Earth. At about 220 light years away, this huge star is an incredible 400 times the diameter of our own sun!

Object 9 - Messier 77 - A Fine and Bright Face-on Galaxy

This galaxy, although noted now by "Messier 77", was actually discovered in 1780 by *Pierre Méchain* who said of it: "Cluster of small stars, which contains some nebulosity, in Cetus & on the parallel of the star Delta, reported of the third magnitude, & which M. Messier estimated to be hardly of the fifth. *M. Méchain* saw this cluster on October 29, 1780 in the form of a nebula."

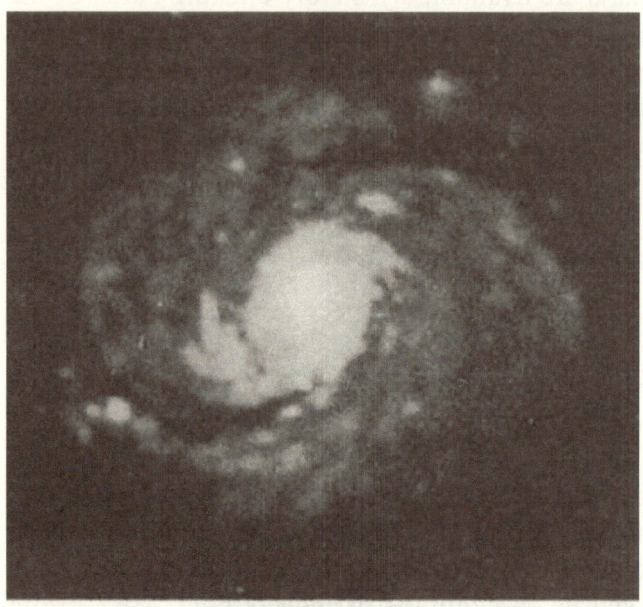

MESSIER 77
A Bright and Beautiful Spiral Galaxy in Cetus

M77 is located only 0.7 degrees ESE from the 4th magnitude star Delta Ceti. The central core of this relatively bright galaxy is about 2 arc minutes and dominates the view of this almost face-on spiral galaxy in 4-inch and larger scopes. A great amount of detail is easily visible with higher magnification in larger instruments operating in very dark sky conditions. Be sure to look for the relatively easy NGC 1055 located about 0.5 deg NNW of M77, visible in the larger scopes as a 3' long edge-on spindle, aligned about east to west, of about mag 10.6. Also note that 11th-mag NGC 1073 is about 1 deg NNE of M77, a face-on galaxy which appears 5' diameter, with a prominent 2x1' bar elongated at about position angle 60 degrees (just north of due east).

With larger telescopes Messier 77 demonstrates a wealth of mottled detail within the limits of the "glow" first seen. During very dark nights and using a magnification of about 140x, a 5 to 8 inch will show considerable irregularity, all of it being the brighter portions of the spiral arms and star conglomerations which can plainly be seen in the sketch shown on the previous page. Note also that the brighter star in seen the lower left portion of galaxy in this (actually a foreground object, NOT part of the galaxy) I have seen easily in 8-inch and larger instruments.

Object 10 - A Faint but Nice Spiral Galaxy - NGC 157
With some discrepancy as to its actual visual magnitude, ngc157 can be seen steadily in the 5-inch on a very dark night. No detail will be seen in this nor the larger telescopes, but it will appear as a very small and uniform glow about 3' arc across. At magnitude 11.2, it is possible to "barely" see this faint face-on galaxy in a 3-inch if the skies are very dark. In all scopes use about 40x per inch aperture for best view.

OBJECT 11 - A Fairly Bright and Large Spiral Galaxy - NGC 247
With a visual magnitude given as 8.9, one might think that this very large (as we see it) galaxy might be very easy in even a 2.5 inch telescope, but this is not the case. Remember that STELLAR

MAGNITUDE is that of a "point source" with the total magnitude compressed into an infinitely small source. Thus a 8th magnitude star is "brighter" visually than an 8th magnitude galaxy, or EXTENDED MAGNITUDE, in which the brightness is "spread out" over the entire surface area. In the case of large ngc247, this means that the 8.9 magnitude is and "integrated magnitude" spread out over an 18.4 x 4.5 minute arc area! Its width is wider than the extent of the full moon! At any rate, this very faint and large object appears as a cloudy "glow" just barely visible in the lowest power you can muster in a 3-inch on the very darkest of nights. If there is any extraneous light or moonlight, you will totally miss it.....in an 8-inch it requires the widest field and lowest power eyepiece you have. There is no detail visible other than merely a very elongated glow of light, like a tiny lone cloud seen through the eyepiece. All the stars you will see at this low power are stars of our own galaxy - foreground objects.

OBJECT 12 - An Elliptical Galaxy - NGC 720
Here is a modest change of pace, being an elliptical galaxy (one with no spiral structure nor traces of the dark dust and obscuring gases common for spiral galaxies) of magnitude 10.5. It is very tiny, yet CAN be seen in small telescopes as a very faint and small "fuzzy star." At 2.1' arc, this very distant galaxy gets better with aperture, but - as characteristic with most ellipticals - shows no detail in any telescopes visually and very little photographically even with the world's giant telescopes.

OBJECT 13 - A Barred Spiral - NGC 937 (with notes on NGC 1073)

There are at least two good barred spirals (appear to have a central bar radiating on each side of the nucleus of a galaxy which appears to "connect" bright spiral arms) in Cetus within reach of larger amateur telescopes: ngc1073 (magnitude 11.4, see listing above) and this one, ngc936. At magnitude 10.7 and fairly large (3.3' arc) this one should be visible in 3-4 inch telescopes, but with no detail. I have distinctly seen both bars in the ETX 125 on a very fine inky-black night with a magnification of about 160x. These bars are distinct, but very faint in the 8" scope. This is a very nice object in really large (16" and above) telescopes. Observers with the larger scopes should check out both these barred spirals and compare; I have found the bars much easier on the fainter galaxy, almost as though the nucleus is an entire bar itself, with definite hints of two major spiral arms at the end of each!

OBJECT 14 - A Very Fine Planetary Nebula with Central Star - NGC 246

Here is a very small (6' arc diameter) but fairly bright planetary nebula that deserves observing by all amateurs. It CAN be seen in small telescopes, but is very tiny and indistinct and can easily be mistaken for a faint blurry star. However, as seen in the Palomar 200" photograph below, there is a wealth of detail within this interesting shell of gas left enclosing a star which exploded in the prehistoric past.

NGC 246
A Great Planetary Nebula With
8th Magnitude Central Star
PHOTO COURTESY PALOMAR OBSERVATORY

NGC 246 is magnitude 8.5 which is very bright for a deep sky object; dark skies and larger apertures will show a wealth of very elusive and intriguing detail worthy of spending a little time here. The central star can plainly be seen in a 6-inch and larger telescope at powers of around 150x and above if the sky is dark. The star's magnitude of 11.3 is deceiving, since it can easily be overlooked from the intertwined loops of nebulosity that may obscure it visually. A good test for the limiting magnitude of your telescope is the faint 14th magnitude star seen at the lower right of the planetary nebula, just on its "edge."

WANDERING ABOUT....YOUR NEW "USER OBJECT" IN CETUS

Rarely do I list a double or multiple star as an "honored candidate" for our USER OBJECT library. Rather, it is normally some freak or conversation object of interest for a star party or public viewing night...something out of the ordinary. However, this Cetus USER OBJECT is merely a double star, an one that is NOT for smaller telescopes. Nonetheless, I challenge Questar users (even 6-inch telescope users!) to locate the companion to "58 CETI" and subsequently log this into their User Objects on the sky program to challenge friends and other observers to find both stars!

58 Ceti is a relatively bright (6.9 magnitude) whitish star with a VERY faint 12th magnitude companion just EAST of DUE NORTH of the brighter star. It separation is ONLY 2.0" arc.....so this is most definitely a challenge! Are YOU up to it? Remember, on such stars the brightness of the primary star frequently overshadows attaining a glimpse at the fainter member! See my article discussing such difficulty when attempting to spot the companion star to the brightest star of the sky, Sirius, on the ASO website: www.arksky.org .

On AutoStar or your PC program, go to: "Select/Object [enter]...." scroll down to "User Object" [enter]. Now enter the coordinates given above for "runaway", using the number keys on AutoStar. After entering the coordinates and pressing "Enter" yet again, scroll down one and you can list the magnitude of the object as "7"[Enter].

Remember that although all telescope sky programs work logically the same way, each one varies slightly in the exact keystrokes or touches necessary to bring up objects from the menus.

A good example is the star MENKAR here in Cetus. On Autostar, it is a simple matter of using the primary menu under OBJECTS, the pulling up (scrolling to) STARS. Under that heading, the sub-categories of Stars leads you to a selection option:

Named (alphabetical)
SAO Catalog
Double
Variable
Etc.

However, most PC programs are far more inclusive and contain the Hipparchus number and other primary reference catalogs as options.

Typically on most PC programs, there is a GO TO tab under which all sub-categories pop up, like Deep Sky, Stars, Comets, etc. Be sure to carefully read the HELP section on your sky program or App.

*   *   *

"His head is made of stars, but not yet arranged into constellations....."
Elias Cantetti

Chapter 11

COMA BERENICES

The Immortalized Amber Tresses of Ptolemy's
Queen Enshrined in the Heavens

This was my "lucky" thirteenth Constellation Guide
written some fifteen years ago, "GO TO COMA
BERENICES". We briefly touched on this
beautiful and clearly conspicuous constellation in
previous constellation guides, but this small
constellations - marked clearly as a sprinkling of
fairly bright stars forming one of the most beautiful
naked eye clusters - really warrants a "GO TO" tour
of its own.

Coma Berenices gets its name *"Berenices' Hair"*
from the Greek legend concerning the Egyptian
leader *Ptolemy III* (no relation to Ptolemy of
astronomy fame) who went to a war so intense that
his return was unlikely. As a sacrifice for his safe
return, his Queen - Berenices the Second - offered
her fiery red (or in some literary references
"golden") hair....considered to be the finest and
most envious locks in all of Egypt and surrounding
lands....

She placed the shorn tresses at the *Temple of
Aphrodite* in the city of Zephyrium to appease the
gods for his safety in battle, only one day the hair
turned up MISSING after Ptolemy's safe return!
Quick thinking in these times by Royal priests and
attendants would allow one to keep his or her head
in such a situation and the Egyptian astronomer to
the King, *"Conon"*, explained that the gods were SO
pleased with the deep red and glowing hair of

Berenices that they placed it in the skies above so that all could enjoy its beauty.

So in addition to a pile of hair, Coma Berenices is the "*Gateway of the Galaxies*" from our viewing platform of Earth. For it sets a target through which our line of sight is aimed directly at the richest concentration of galaxies visible to mankind. Large photographic telescopes can detect literally thousands of galaxies in one time exposure photograph.....limited ONLY by the film's ability to detect light. Indeed, all the way to the limiting magnitude of the telescope, the film the electronic sensors, galaxies as far as the mind can imagine fill the sky in the direction of Coma Berenices.

Many of these galaxies are visible in our telescopes, and the larger the aperture, the more galaxies obviously can be seen. It is quite a challenge in and of its own for an observer to attempt to log ALL galaxies visible within his or her telescope within this constellation's very limited borders.

BONUS: IN ADDITION to our regular "GO TO" TOUR for this galaxy field (and, there WILL be a couple of objects thrown in that are NOT galaxies for some relief...) I am providing a complete list of the **"major" observable galaxies** by NGC (New General Catalog) listings as well......all that are provided can be found in chronological order under "Object/Deep Sky/NGC....." [then scroll to select the appropriate number]. In addition, I am providing: [your sky program procedure may vary]

1) **NGC** number (and Messier # if applicable);

2) R.A. and DEC. coordinates if you chose to enter yourself;

3) visual magnitude;

4) overall dimensions, in minutes (') arc (remember, the moon's disk is 30' arc, Jupiter is just less than 1' arc)

5) the galaxy TYPE (either "S" for spiral, "E" for elliptical, or "Pec" for irregular or peculiar;

6) comments on scope requirements and expected views.

In all, we will have an opportunity to view as many as 30 galaxies within this one tiny constellation!

This gorgeous naked eye cluster is easily found by extending an imaginary line from POLARIS through ALIOTH (in Ursa Major) southward until it intersects with this very bright sprinkling of "star dust", all stars appearing very much the same magnitude, just within reach of the naked eye. Although no galaxies can be seen with them, binoculars offer a spectacular view of this cluster.....also known as "**Mel 111.**" The center coordinates of Mel 111 are: R.A. 12h 23m; DEC + 26 degrees 24m.

The reason for its conspicuousness is that the Coma cluster is only 250 light years away.....this is about half the distance to the famed Pleiades and just a bit farther than our closest star cluster: the Hyades, both in Taurus the Bull. Most stars of this group fall in magnitude between 6.0 and 10.2, the brightest of which are visible to the eye. In terms of star numbers, it has only about 25% as many stars as does the Pleiades and seems to be about the same age as that cluster, but a much younger star group than the Hyades or the other conspicuous naked eye galactic cluster....Praesepe, or the *"beehive cluster"* in Cancer.

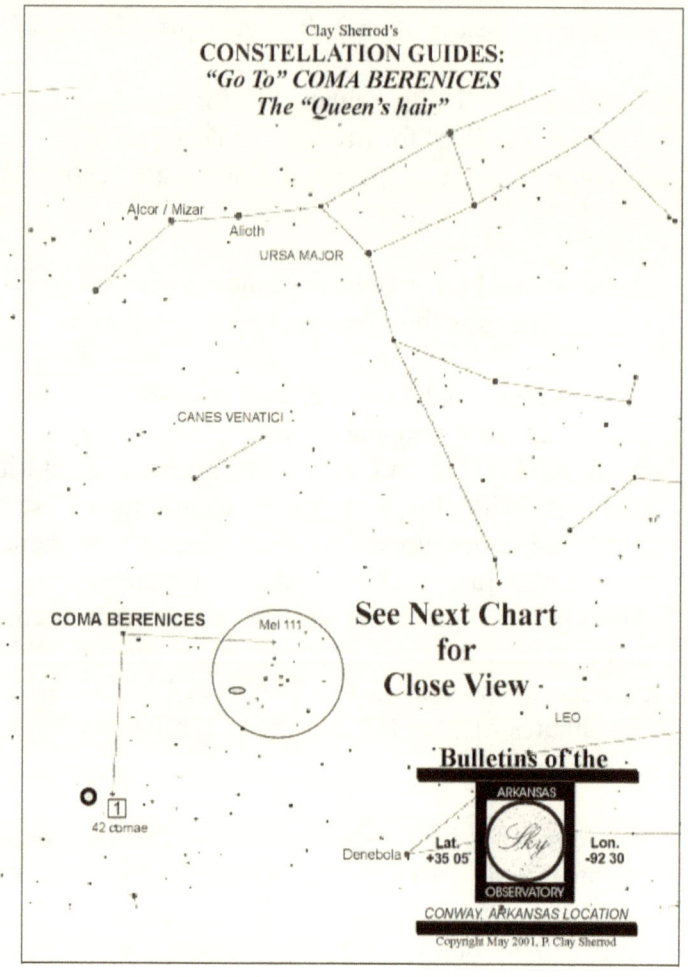

Clay Sherrod's
CONSTELLATION GUIDES:
"Go To" COMA BERENICES
The "Queen's hair"

A finder chart for locating many of the GO TO objects in the constellation of Coma Berenices; if using a computer planetarium program, you are encouraged to plot the objects on your screen for higher resolution than this chart provides.

*   *   *

You will begin your "GO TO" journey into Berenice's Hair (we really hope she has washed it....) a bit differently than with most of our "GO TO" constellations which always start with the brightest star in that star group. With Coma Berenices you will begin your GO TO with the command "Object/Constellation/Coma Berenices"! Now, that's different....it will take you to the brightest star of the cluster, and hence the constellation itself.

Each GO TO object is discussed for your telescope regarding the type of conditions necessary for you to view it optimally for discern the very faintest details....one double star challenge for all telescopes.....magnifications and aperture necessary for most objects, and much, much more. This is YOUR complete GUIDE to get you on your way to exploring this large and interesting constellation. For a complete listing an descriptions of the multitudes of galaxies visible in this constellation and surrounding regions, consult a good handbook, such as the "*Burnham's Celestial Handbook*," Vol. 2 for a very comprehensive list of locations, magnitudes and angular separations of these wonderful deep sky objects. There are many galaxies for EVERY telescope size and type.

Indeed, this TOUR will give many of you your first opportunity to view something BESIDES the typical "spiral" galaxy that is so popular....the roundish "elliptical" galaxies abound in this region of our skies, and the irregular galaxies (much like the bright **Messier 82** in Ursa Major - (see my Constellation Guide - Ursa Major under GUIDES on this website) offer a glimpse at the evolutionary cycle of the Universe's most magnificent and mysterious objects.

Use the attached star chart and the following Guide as an excellent reference for your next star party itinerary, or a beginning for further study into the thousands of objects visible in this part of the sky. Merely click on the image above, save your chart to file, open the file and resize to fit your page and PRINT! Truly these extensive Constellation Study Guides will most definitely put your computer to work for you in the most efficient and enjoyable way possible! As an example of how these tours can be loaded successfully into your library as a custom Tour, visit those for Leo and Bootes which have been compiled and offered through this ASO site under the GUIDES/Constellations.

We hope you enjoy these comprehensive GUIDES to touring the constellations via your AutoStar or PC sky program and its computer-driven telescope. Each new installment is complete with diagrams, charts and illustrations that you will find nowhere else. Please let us hear YOUR feedback and your observations of each and every constellation after YOU have toured its vast reaches of our skies!

YOUR COMA BERENICES CONCISE DIRECTORY OF INTERESTING OBJECTS –

Although there is a "smattering" of other objects in our Coma Berenices "GO TO" TOUR, you will see that this is most definitely a compendium of galaxies.....much like the "GO TO" TOUR for the constellation Ophiuchus posted here on this website under GUIDES/Constellations was for the globular clusters! Indeed, we will find two (2) globulars even in THIS review as well.....both very nice and extremely interesting for comparison!

I have chosen the finest 11 objects in this COMA BERENICES "GO TO" TOUR; as with all GUIDES, all objects listed below will be visible in all telescopes (some naked eye) from the 3-inch through 6-inch; of course larger apertures may "show" an object a bit closer and "better," but frequently a wide field and low power view is more desirable than aperture. Once your eyes are fully dark adapted, you will even be able to see much more detail and the true expanse of the Coma cluster - **Mel 111**. Nearly twice the number of stars will be visible when the constellation is nearly overhead for northern hemisphere observers (the constellation is nearly overhead - or "culminates" - at midnight during the last week of every March.

As with all of the "GO TO" TOUR constellation lists, I recommend a good star atlas and/or chart which will list all the finest objects, constellation-by-constellation. One very handy reference guide is the *PETERSON FIELD GUIDE TO THE STARS AND PLANETS*, which features complete lists with declinations, right ascensions, magnitudes, and all pertinent information for you to expand your observing horizons beyond this brief guide.

Use the attached star chart (above) and the following Guide as an excellent reference for your next star party itinerary, or a beginning for further study into the thousands of objects visible in this part of the sky.

Following is the concise object list for your "GO TO" TOUR of Coma Berenices; you may wish to find the majority of the objects from the PC Library (for example, you can easily go to the "Black Eye Galaxy" if you pull up "Object/Deep Sky/Messier Object/..type in '64'...." and then press "Enter", followed by "GO TO" to access an incredibly interesting galaxy that appears to have a "hole" in it! In addition, if you have never tried it, go to "Object/Deep Sky/Named...." and scroll to "Black Eye Galaxy" that is in among your computer library....that will get you there just as quickly! LIKEWISE....you can access the same object via AutoStar or other keypad by going to: "Object/Deep Sky/NGC....." and typing in "4382" as you will do for the NGC listing that follows outside of our normal tour for this constellation. On the other hand, if you want to experiment and become a "better computer user" try entering the exact R.A. and DEC coordinates of that object as described above after holding down the MODE key. You will find the accuracy of entered GO TO's to be somewhat less than those stored in planetarium program, but the capability of acquiring unlisted objects is fantastic!

OBJECT 1: constellation - COMA BERENICES - (alpha Com) R.A. 13h 08' / DEC + 17 48 - Magnitude: 4.2
OBJECT 2: double star test - 37 Comae - R.A. 12h 58' / DEC + 31 03 - Mags. 5 & 13 - Perfect test for 3-inch scope!
OBJECTS 3: wonderful pair of Globulars! - Messier 53 (ngc5024) & ngc5053 - R.A. 13h 11' / DEC + 18 26 - Mag. 8

OBJECT 4: "*BLACK EYE GALAXY*" - Messier 64 (ngc4826) - R.A. 12h 54m' / DEC + 21 57 - Mag. 8 - NICE!

OBJECT 5: Elliptical galaxy - Messier 85 (ngc4382) - R.A. 12h 23' / DEC + 18 28 - Mag. 10.5 - not easy....

OBJECT 6: Tilted spiral galaxy - Messier 88 (ngc4501) - R.A. 12h 30m / Dec + 14 42 - Mag 10.2 - Elongated shape

OBJECT 7: Spiral galaxy - Messier 91 (ngc4571) - R.A. 12h 34m / DEC + 14 28 - the "missing" Messier object!

OBJECT 8: Edge-on spiral galaxy - Messier 98 (ngc4192) - R.A. 12h 11m / DEC + 15 11 - VERY large, dim

OBJECT 9: "Pinwheel Galaxy" - Messier 99 (ngc4254) - R.A. 12h 16m / DEC + 14 42 - Face-on, very bright!

OBJECT 10: Spiral galaxy - Messier 100 (ngc4321) - R.A. 12h 20m / DEC + 16 06 – VERY large, face on – NICE

OBJECT 11: Fantastic edge-on galaxy! - NGC 4565 - R.A. 12h 34m / DEC + 26 16 - "The Needle Galaxy!" WOW

OBJECTS 12 THROUGH 34 - NGC GALAXIES IN COMA BERENICES (other than those listed above) complete listing of "ngc" galaxies in Coma Berenices in order of RIGHT ASCENSION (and NGC # order):

NOTE: These NGC galaxies are NOT detailed in the following "Visual Guide" as are those Objects 1-11 listed above...use the abbreviated descriptions as they follow the order:

NGC# / R.A. / DEC / MAGNITUDE / SIZE (in minutes arc -'-) / GALAXY TYPE (S=spiral, E=ellipt;, P=Irr.)

ngc4136 / 12 06 / +30 12 / 12.0 / 3.4 x 2.8 - spiral, face-on, 8-inch+!

ngc4203 / 12 13 / +33 29 / 11.0 / 1.8 x 1.5 - elliptical, tough for 6-8 inch scopes, round blob

ngc4212 / 12 13 / +14 11 / 11.7 / 2.3 x 1.3 - tilted spiral, tough for 3-inch..elongated with 6-inch +

ngc4251 / 12 16 / +28 27 / 10.2 / 2.3 x 0.8 - edge-on spiral, cigar shaped, test for 2.5"; good in 8-inch+

ngc4274 / 12 17 / +29 53 / 10.8 / 6.7 x 1.3 - barred spiral, nice in 8-inch; tough in smaller

ngc4278 / 12 18 / +29 34 / 10.3 / 1.4 x 1.3 – very tiny, starlike; very tough with 3-inch, a spot with larger scopes

ngc4293 / 12 19 / +18 40 / 11.5 / 4.6 x 1.6 - P: irregular very elongated; only seen in 8-inch+

ngc4298 / 12 19 / +14 53 / 11.8 / 2.2 x 1.1 - tilted spiral - VERY hard....may not be seen in 8-inch

ngc4314 / 12 20 / +30 10 / 10.8 / 3.1 x 2.9 - large barred spiral, can be seen in 4-inch but tough

ngc4350 / 12 21 / +16 58 / 11.6 / 1.8 x 0.5 - extremely small, elongated elliptical. A test for the8-inch

ngc4394 / 12 23 / +18 29 / 11.2 / 2.3 x 2.3 - barred spiral, a test for the 5-inch, neat galaxy

ngc4414 / 12 24 / +31 30 / 9.7 / 3.2 x 1.5 – tilted spiral, can be seen steadily in 3-inch; good in larger

ngc4419 / 12 24 / +15 19 / 11.4 / 2.2 x 0.6 - very hard elliptical, very small, starlike in 8-inch

ngc4448 / 12 26 / +28 54 / 11.4 / 2.8 x 1.0 - tilted spiral, seen only in 6-inch+ and larger, cigar shaped

ngc4450 / 12 26 / +17 21 / 10.0 / 3.0 x 2.5 - spiral, fairly large and oval in 3-inch; good in larger scopes

ngc4459 / 12 27 / +14 15 / 10.9 / 1.2 x 1.0 - tiny elliptical, mistaken for star in 4-inch; seen okay in larger

ngc4477 / 12 28 / +13 55 / 10.7 / 2.4 x 2.2 - face-on spiral, visible in 5-inch

ngc4494 / 12 29 / +26 03 / 9.6 / 1.3 x 1.2 - very small elliptical, starlike - visible in 4-inch and up scopes

ngc4548 / 12 33 / +14 46 / 10.8 / 3.7 x 3.2 - tilted spiral, visible in 6-inch as faint elongated glow

ngc4559 / 12 34 / +28 14 / 10.6 / 11.0 x 4.5 - tilted spiral - huge, but VERY dim in 6-8 inch scopes

ngc4689 / 12 45 / +14 01 / 11.8 / 2.4 x 1.8 - tilted spiral - small and very, very dim..tough in even 8"

ngc4725 / 12 48 / +25 46 / 8.9 / 10.0 x 5.5 - very large and easy in 3-inch; may be seen in small APO!

....AND NOW ON WITH THE SHOW!! (refer to the Coma Berenices close-up chart shown below for all the objects described in detail on the "Guide")

A VISUAL GUIDE TO OUR DEEP SKY OBJECTS IN COMA BERENICES

Object 1 - Constellation GO TO - Coma Berenices (42 Comae, alpha Com)

As stated we are staring out this "GO TO" TOUR with a different twice, taking you to the heart of the constellation via the planetarium program's "Constellation Library." This keys in on the brightest star of each constellation, in this case "42 Coma Bernices", at a dim magnitude 4.23. There is nothing particularly interesting about this particular star, but I wanted to provide the opportunity for you to use your GO TO / Constellation key at least once in your computer!

However, this IS a very difficult double star that theoretically can be resolved with the 8-inch and even perhaps the 6-inch under very high magnification and steady seeing. If you CAN resolve it....you deserve the night off. I cannot split this one with my 8-inch. Both stars of this pair are magnitude "5" and presently are UNDER the maximum separation of about 0.9' arc. This pair is a relatively close 65 light years distant, and both are about 3 times the size of our own sun, although very similar in all respects

Clay Sherrod's
CONSTELLATION GUIDES:
"Go To" COMA BERENICES
Enlarged View of Cluster/Constellation

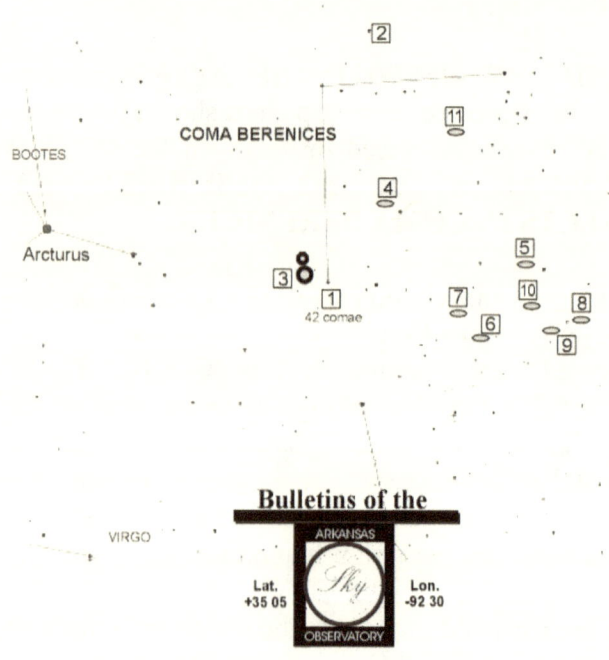

<u>Object 2</u> - 37 Coma Berenices - A Very Tough Double for the 6-8 inch scopes

This is an easy star to find, immediately above and just west of Beta Comae, as shown in the close-up chart. However, this is an extremely difficult double star....the main star is magnitude 5.0 while the secondary star is ONLY magnitude 13; it is "possible" to see this star in the 4-inch nearly due north of the brighter star, but more likely seen in the 6-inch and larger scopes. A 13th magnitude star is visible in both scopes under the darkest of sky conditions, but this one is only 5.2" arc from the 5th magnitude primary star, so its glare may prevent you from seeing it. I would be interested from 8-inch scope users if this star is visible to them....I can detect it, but ONLY because I know it's there! This may be the toughest double star you have had yet!

<u>Objects 3</u> - A Wonderful Pair of Globular Clusters! - Messier 53 and NGC 5053

This pair of globular clusters is separated in your field of view by only one degree, and the differences between these two beautiful objects is very interesting. You can spend much time just studying and noting the subtle peculiarities when comparing. Messier 53 is a very fine globular, bright and large with a "typical concentration" of stars, densely packed into the center. It reminds me of a somewhat smaller and fainter Messier 5 in *Serpens* (see: *Ophiuchus* here in the GUIDES.

Although just on limit of star resolution in the 3-inch, expect some partial resolution with this scope along the perimeter of the larger Messier 53. The stars are fairly bright for a globular - 11th and 12th magnitude. But with a total brightness of only 8.7, this is a bit of a difficult object. Photographically, the globular has a surprisingly BRIGHT magnitude of only 3.7! In the 6-inch and larger scopes this is an outstanding sight with glittering spattering of "star dust" seen all the way to the core of this "half-moon-sized" object. Look only 1 degree southeast of M-53 (use the 40mm eyepiece with the smaller telescopes....well what can I say: they'll both be in that super wide field of view, but with ngc5053 right on the verge of being too faint to see), and you will find one of my favorite globular clusters, the peculiar NGC 5053. The jury is still out as to whether this is actually a globular cluster or perhaps a very DENSE galactic cluster, like Messier 11 in *Scutum* (see *Aquila*). This is a beautiful but very difficult sight, only about magnitude 10.2 but still clearly visible in the 6-8 inch scopes. Compare this loose globular to M-53 and just look at the difference in the concentration of stars toward the center! Interesting, both objects are about the same distance and the fainter is actually 10,000 light years CLOSER than the brighter M-53, at "only" 55,000 light years. This makes both of them among the most distant of all globular clusters in this portion of the sky! Expect little or no resolution of the individual stars even in the 8" telescope.

Object 4 - The "BLACK EYE GALAXY" - Messier 64

Here is one you can dial up by NAME under OBJECTS / DEEP SKY / NAMED.... and scroll until you get to "Black Eye Galaxy", or merely type in Messier "64" under that listing. This is a very fine object in all scopes, appearing as an elongated bright oval shape, fairly large, in the smaller scopes; in the 6-inch you will begin to see the full extent of this nice galaxy, magnitude 8. However, it takes the 8-inch and a very dark night to begin to detect the "black eye" so don't expect too much out of this from its label! This is one "deep sky exception" in that using higher than usual magnifications will actually enhance your chances of glimpsing this interesting dark marking against the bright center of M-64. With the larger scope, the dark portion is clearly visible and offset greatly to one "hemisphere" of the core of the galaxy; in addition, I was able to detect ONE dark spiral hint just opposite that dark matter (see the photography, courtesy U.S. Naval Observatory, below) from the core of the galaxy. One thing that is VERY noticeable with all scopes the very bright star-like "nucleus" of this galaxy. This is an outstandingly bright and large (7.5 x 3.5" arc) galaxy with a beautiful oval shape.

Messier 64
The "Black Eye Galaxy"
U.S. Naval Observatory Photo

<u>Object 5</u> - Our First "GO TO" Elliptical Galaxy - Messier 85

At magnitude 10.2 or so, this galaxy is a bit easier than that brightness would indicate; however don't expect anything but a roundish 3' arc by 2' arc round glow. In the ETX 125 I can distinctly make out a very star-like nucleus of the galaxy. This elliptical galaxy CAN be seen in a small APO refractor, but will appear very much like a very faint star that is out of focus....it is likely to be missed altogether unless you crank up the magnification somewhat. This object is an incredible 41 million light years away and contains enough mass that would require over 100 billion suns!

For an 8-inch, look for the VERY faint galaxy NGC in the same medium power field of view, 8' arc to the east. This fainter galaxy, at visual magnitude 11.7 and very difficult, is a barred spiral which should show some elongation as a very small oval.

<u>Object 6</u> - Messier 88 - A Fine "Tilted" Spiral Galaxy

Some galaxies present their beauty to us face on, like the famous "Whirlpool Galaxy" in *Canes Venatici* (see Constellation Guides here on this website) or perhaps EDGE-on, like the wonderful "Needle Galaxy" also in Coma Berenices (discussed following). In the case of Messier 88, we are looking at something in between: a spiral galaxy that is tilted toward our line of sight by about 30 degrees, making it appear much elongated and characteristically "pointy" on each end. It is a large galaxy, almost 6' x 3' in size and appears very nice in even the 3-inch scope, much brighter than its given 10.5 magnitude; use medium to medium-high powers on the 6 to 8-inch scopes and you likely can glimpse some spiral structure if viewing from a VERY dark sky. The size of this galaxy makes it a good, but difficult, target for smaller telescopes. Interestingly, an equally nice galaxy - NGC 4571 - is located only one degree east and a bit to the south of M-88 and is "thought" to be the one missing Messier Object as discussed below.

Object 7 - Messier's "**Missing Galaxy**," – Messier 91 (??)...or was it really a comet?

Charles Messier did not start out his famed astronomy career desiring to catalog "faint fuzzies" in the sky; indeed, he did not even LIKE them....so much so that they were getting in his way of fame by seeking out new comets on a regular basis from his observatory in France. He would no sooner think that he had discovered the wonders of a new interloping comet that to find out that it was merely a faint fuzzy that was always there....only nobody had ever bothered to write down the positions of these obstacles! So Messier cataloged as many objects as possible so that he could provide a quick and reliable reference to objects and dismiss them as comets should he come across them again....so were obvious, like the Praesepe (M-44) and the Orion Nebula (M-42), but other actually LOOKED like distant comets before they form the characteristic tail. In March, 1781, Messier recorded an object with the description he used so many times: "....a nebula without a star...fainter than -90", referring to his entry for now-Messier 90. His cataloged position was: RA 12h 35.0m, DEC 14 degrees 02 minutes, only there is NO object at that position! Perhaps it was, indeed, a comet and even Messier was fooled! However, there IS an object at

RA 12h 34.3m, DEC +14 degrees 28 minutes that is likely his object. This is a face-on spiral galaxy, only magnitude 11.2 visually and only about 2.5' x 2.2' diameter. This object is fairly conspicuous in both the 6 and 8-inch telescopes for such a dim magnitude designation, and is an easy target even in a 3-inch.

Object 8 - Another Spiral Galaxy - Messier 98
After fighting to see any detail in Messier 91, turn your attention to the VERY large and fairly bright Messier 98, only 1/2 degree west of the 5th magnitude star 6 Comae. The closeness of this star adds to the ease of locating the object and provides a super impression of "3-D" between the star and galaxy. Messier also noted this was a "...nebula without star...." (yawn), but failed to mention it is huge in the telescope, measuring 8' arc long and 2' arc wide, making it a very interesting cigar-shaped object in medium power. It is magnitude 10.7, but can be easily seen in small APO refractors with medium-high power. Larger telescopes might reveal that it is truly a nearly edge-on galaxy, with some detail around its perimeter. It is a fine sight in the 6-8 inch scopes at about 100x or so. Look in the same 1-degree field of a low power wide field eyepiece in those scopes for Messier 99 to the east and slightly north of M-98.

Object 9 - Messier 99 - the "*Pinwheel Galaxy*" (but not if you believe that M-33 is the "real pinwheel"
This nice galaxy is frequently mislabeled the "Pinwheel Galaxy" in astronomy references, but that honor goes to the famous and large **Messier 33** in Triangulum. Nonetheless, Messier 99 is a true "pinwheel" shape and this spiral structure can be

detected on a very dark night at about 125x in an 8-inch and suspected in a 6-inch at about 90x. It appears as a roundish glow in the 3-inch and smaller scopes with a very bright concentration - almost star-like - at the very center. At only 4' arc and round, the galaxy WILL show some of its brighter arm structure to very large amateur telescopes, and fairly impressively so! With a mass of over 50 billion suns, when you gaze at this distant galaxy, you are looking at light that left those suns some 60 MILLION YEARS AGO, just now getting through the focal system of your telescope!

Object 10 - A Wonderful Spiral Galaxy for all Scopes! - Messier 100

I suspect Messier was getting somewhat weary of recording "faint fuzzies" by the time he got to his 100th listing, the beautiful face-on spiral in Coma Berenices. I am sure that his description "....difficult to recognize....feeble light" was a badly needed drift from the typical "....nebula with no star." For M-100 is a spectacular objects in even the smallest APO refractors and larger telescopes; although visible clearly in the small refractors, it is devoid of detail in those scopes. With the 6-inch, I was clearly able to observe the spiral structure of this beautiful galaxy (as shown in the accompanying photograph, courtesy the Mount Palomar 200" telescope) with the 8-inch when the galaxy was nearly directly overhead my observatory at dark Petit Jean Mountain. This is a truly huge galaxy, comparable in size and mass to the Great Andromeda Galaxy, and measuring in diameter nearly 120,000 light years! It is only magnitude 10, but is clearly distinguishable and a definite "must-see.

Messier 100
The "Pinwheel Galaxy"
(see text)
Mount Palomar 200" photo

<u>Object 11</u> - My Favorite Galaxy - NGC 4565 - the
"Needle" Edge-on Galaxy

You've heard of the "space needle?" Well, here is
the real deal. If the sight of this galaxy does NOT
give you an adrenalin rush then you better schedule
your monthly physical right away. This is the
classic of all edge-on galaxies and by far the most
interesting (with Messier 82 in Ursa Major a close
second) of all galaxies to view in amateur

instruments. This long needle will literally FILL your eyepiece from point to point across a NE to SW direction at about 100x; I have found that magnification is ideal for ALL scopes. Lower magnification loses some of the nice contrast necessary to see the full extent of the object, and higher powers merely blur the sharp detail which "may" be seen. On a very dark night, you might see this galaxy much as it appears in fantastic photograph from the 200" Palomar telescope below:

NGC 4565
The "Needle Galaxy"
Mount Palomar 200" photo

In the 6 and 8-inch scopes, both at about 100x I can clearly make out the dark "lane" of the very edge of this razor-sharp image. The galaxy tip-to-tip literally touches the edges of my fields of view! You can clearly make out the "lens-like" hub of the galaxy. All in all, it looks pretty much just like the photograph seen, only not as bright as captured on film. NOTE that the tiny star seen in this photo just to the left of the galaxy's hub can clearly be seen in a six-inch at 100x when the air is very steady. It can be held unmistakably in the 8-inch and larger scopes. In smaller scopes, locate the field first and move the barrel of the scope slightly to reveal the faint sliver of light; then, once centered, increase the magnification up to about 50 x or 60x for the best views. This galaxy measure 15' arc long.....half the size of the moon's disk! Thus, you can appreciate how much is lost at higher magnifications. Although its magnitude is given a visual 10.2, it is really hard to understand how *Charles Messier* missed this one....perhaps it was not enough "....nebula without star..." for him to have taken note. Nonetheless, you should take note...this is one object that you should NEVER do without. It is a showpiece of the sky and great for star parties and telescope cookouts!

WANDERING ABOUT....YOUR NEW "USER OBJECT" IN COMA BERENICES

Since Messier ignored the fantastic sliver of **NGC 4565**, we will not. Although you can access this object through the OBJECT / DEEP SKY / NGC....[enter number] library of your PC program, let's do this object justice and program your

Autostar and key in the coordinates for this "needle" as given above.

On AutoStar or your PC program (use tabs and instructions for USER OBJECTS on various programs), go to: "Select/Object [enter]...." scroll down to "User Object" [enter]. Now enter the coordinates given above for "NGC 4565", using the number keys on the computer. After entering the coordinates and pressing "Enter" yet again, scroll down one and you can list the magnitude of the object as "10"[Enter]. Now go back and mode back to "USER OBJECT / Description...." and enter "needle" or something equally creative....NOW, you can access this wonderful object directly from your own custom computer growing USER OBJECT LIBRARY! Messier would have been envious of our abilities to eliminate all these faint fuzzies in h is comet searches by merely consulting a computer handbox full of tens of thousands of such objects!

*   *   *

"Every luminary in the constellation of human greatness, like the stars, comes out in the darkness to shine with the reflected light of God....."
                                      Mary Baker Eddy

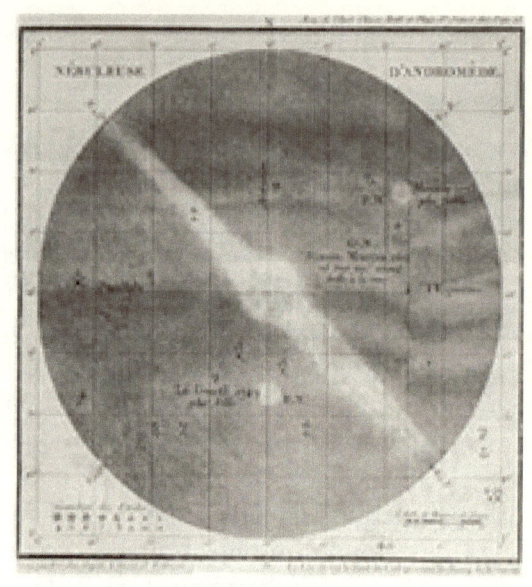

An example of the original Messier Catalog
By Charles Messier (below)

Chapter 12

CYGNUS

(and surrounding constellations of Delphinus, Sagitta, and Vulpecula)

Scoping in on the Striking Celestial Swan

This *Constellation* Guide, "GO TO CYGNUS" of this series takes us RIGHT SMACK in the middle of the beautiful **Milky Way Galaxy** as we explore the famous "North American Nebula" and observe the very dark "lane" of obscuring dust in the direction of our neighboring galactic arm of stars. In Cygnus we marvel at the array of wonderful deep sky objects, dense clouds of gravitationally-bound stars, dark obscuring and light-absorbing space dust and wonders beyond imagination. In **Cygnus**, tiny **Delphinus** (the "dolphin") **Sagitta** ("the archer's arrow") and **Vulpecula** (the "Fox") the star clouds are phenomenal and worthy of very wide field and slow scans throughout the constellation. It is pocked with very bright and contrastingly interesting colored stars, clusters of all types, both bright and dark nebulae and one of the most exciting planetary nebulae - "The *Dumbbell Nebula*" (Messier 27 in Vulpecula) that you will ever observe.

In addition be prepared for perhaps the most striking double star, *Albireo*, that you will ever observe in any 3-inch and larger telescopes. No matter how many times you see this star....you will keep coming back to it. It is truly a showcase object for a star party or your next neighborhood barbecue sky session! But be sure to note that Albireo is by far not the only interesting - even spectacular - double star in Cygnus, Delphinus or Sagitta....there are many, many more worth pursuing. There are few such objects on this GUIDE, but also be sure to consult your sky library or your scope handbox under for more GO TO objects, and find a good atlas or PC sky program for the hundreds more not listed on your hand controller.

You will begin your "GO TO" journey into Cygnus via the bright star "*Deneb*", a nice bright yellow-white **supergiant** star that is commonly referred to in the "asterism" known as the *SUMMER TRIANGLE* (below), a nice wide shape bounded by the bright summer stars Deneb (Cygnus), Vega (Lyra) and Altair (Aquila).

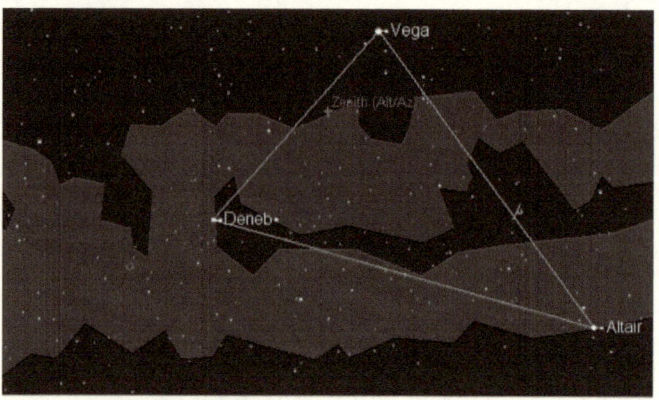

Courtesy Starry Night

The approximate locations of our lesser constellations - **Delphinus**, **Sagitta** and **Vulpecula** - can be clearly noted from this and the following regional sky chart.

Each GO TO object is discussed for your telescope regarding the type of conditions necessary for you to view it optimally for discern the very faintest details....double star challenges for each size telescope .....magnifications and aperture necessary for most objects, and much, much more. This is YOUR complete guide to pique explore the exciting inner arm of our Milky Way galaxy, its exciting large galactic star clusters, bright emission nebulae and the dark obscuring dust that blocks the otherwise life-threatening radiation from the very center of our galaxy. Use these guides as an excellent reference for your next star party itinerary, or a beginning for further study into the thousands of objects visible in this part of the sky.

Truly these extensive *Constellation* study guides will most definitely put your sky program to work for you in the most efficient and enjoyable way possible!

We hope you enjoy these comprehensive guides to touring the constellations via your sky program and its computer-driven telescope. Each new installment will appear frequently, complete with diagrams, charts and illustrations that you will find nowhere else. Please let us hear YOUR feedback and your observations of each and every constellation after YOU have toured its vast reaches of our skies!

Introduction   -

CYGNUS (pronounced "SIG-nus") is often referred to as the "*Northern Cross*" as its shape denoted by the brighter star members truly do conform to that of a large summer cross overhead. Its counterpart "**Crux**", or the Southern Cross, is a favorite for observers and even the earliest sea-going navigators through the southern hemisphere skies.

We have discussed in other constellation "GO TO" guides that many constellations did NOT look like their namesakes, and actually neither does Cygnus. Thus, I have always found it remarkable and very curious that the constellation throughout many cultures and times has ALWAYS denoted some time of bird in our celestial aviary. To some it was a *crow* (now given to "Corvus"), and a peacock.....a swan....and even a chicken. In earliest Arabian writings, the principal star DENEB was titled AL DHAHNAB....or *"tail of the hen."* It is from this description that the star gets its present name. Still other Arabic labels showed the constellation as a swan, an eagle, a dove....perhaps from earlier Babylonian script. (looks like a buzzard, below)

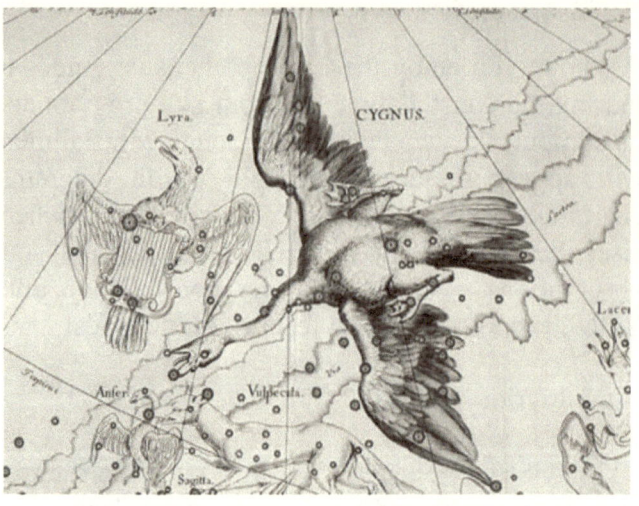

Cygnus as drawn in Hevelius' Uranographia

The entire constellation is worthy of very wide field scrutiny and the small APO refractors are ideal for dark night scanning at their lowest magnifications. For through Cygnus, running the length from the "top" to the "bottom" of the Northern Cross is the magnificent star clouded arm of our Milky Way....but let's NOT stop there.

Interrupting this sky's length is an equal length of DARK MATTER, light-obscuring and energy-absorbing dust that blocks both light and high energy from the center of our Milky Way (nearby and south in Sagittarius). On a dark night with the naked eye, and with the low wide field telescopes, this dark cloud is truly remarkable as it literally BLOCKS the light behind it. It appears as a black streak smeared across the expanse of the bright clouds of the Milky Way.

Wait until Cygnus has risen very high in the sky and recline for some of the most wonderful views you have ever experienced in this magnificent area. A pair of 7 x 50 or 10 x 50 binoculars will reveal MORE than you could possibly explore in a lifetime in this region!

Look for the crafty FOX (*Vulpecula*) chasing the "head" of the swan (*Albireo*), just south of the bottom of the "cross." Sagittarius' errant arrow (*Sagitta*) is flying just under this fox and south also of Cygnus, while the playful Dolphin of the sky (*Delphinus*) leaps just southeast of all this from the watery crests of the bright Milky Way. Although only composed of five primary stars, Delphinus is one of the few constellations that actually LOOKS like what it is supposed to be....a leaping Porpoise.

YOUR CYGNUS / Delphinus / Sagitta / Vulpecula CONCISE DIRECTORY OF OBJECTS -

This group of summer constellations is full of outstanding and curious objects, stars, clusters, bright AND dark nebulae - from naked eye, to wide field, to telescopic - that is was very difficult to select (and limit) the number of objects for our "GO TO" TOUR. I have chosen the finest 14 objects in this CYGNUS and surrounding area to showcase for your "GO TO" TOUR; in this case, all objects listed below will be visible in all telescopes (some naked eye) from the 3-inch to 8-inch telescope; of course larger apertures may "show" an object a bit closer and "better," but frequently a wide field and low power view is more desirable than aperture. This is the case for MANY of these objects since we are looking directly into the very star-cloud-rich areas where wide view is more desirable than aperture.

This is the case for MANY of these objects since we are looking directly into the very star-cloud-rich areas of our Milky Way galaxy. Indeed, I strongly encourage your to step away from the telescope often and scan the beautiful open skies and star fields with a good pair of 7 x 50 or 10 x 50 binoculars. You will be tempted to venture to your "dark sky site" for a full evening of laying back on a blanket and scanning the skies. The many deep sky objects in the Cygnus area even stand out to the dark-adapted naked eye, and the dark and bright star clouds emerge like in no other area.

Clay Sherrod's
CONSTELLATION GUIDES:
"Go To" CYGNUS- The "Celestial Swan"
with discussions of Vulpecula and Sagitta

LACERTA

DRACO

6 13 1
Deneb
9 12 3

"Double-Double"
5 Vega
CYGNUS LYRA
8
M-57
7

4 11
2 Albireo

10 VULPECULA

EQUULEUS

SAGITTA

14 DELPHINIUS

Bulletins of the

ARKANSAS

Lat. Lon.
+35 05 -92 30

OBSERVATORY

Altair
AQUILA

A finder chart for locating many of the GO TO
objects in the constellation of Cygnus; if using a
computer planetarium program, you are encouraged
to plot the objects on your screen for higher
resolution than this chart provides.

*　　　*　　　*

307

Once your eyes are fully dark adapted, you will even be able to see such wonders perhaps as the magnificent and very large "*North American Nebula*" (discussed below), just to the northeast of bright Deneb. Even the binoculars will show dozens of deep sky objects that cannot be fully appreciated in large telescopes with limiting fields of view!

The high northern altitude of Cygnus places it for ideal summertime observing for northern hemisphere skygazers. Rising nearly due east and transiting the meridian only 6 hours later, this constellation affords a full night's observing pleasure and discovery.

As with all of the "GO TO" tour constellation lists, I recommend a good star atlas and/or sky program which will list all the finest objects, constellation-by-constellation. One very handy reference guide is the *PETERSON FIELD GUIDE TO THE STARS AND PLANETS*, which features complete lists with declinations, right ascensions, magnitudes, and all pertinent information for you to expand your observing horizons beyond this brief guide.

The constellation tour Star Chart above will get you started on your journey for this constellation.

Following is the complete 14-object list for your "GO TO" tour of Cygnus; you may wish to find the majority of the objects from the computer Library (for example, you can easily go to the Dumbbell Nebula if you pull up "Object/Deep Sky/Messier Object/..type in '27'...." and then press "Enter", followed by "GO TO" to access my favorite all time globular cluster. On the other hand, if you want to experiment and become a "better computer user" try entering the exact R.A. and DEC coordinates of that object as described above after holding down the MODE key. You will find the accuracy of entered GO TO's to be somewhat less than those stored in your computer, but the capability of acquiring unlisted objects is fantastic!

OBJECT 1:   very bright star - Deneb (alpha Cygni) - R.A. 20h 40' / DEC + 45 06 - Magnitude: 1.2, super giant star

OBJECT 2: *SKY'S BEST DOUBLE* - Albireo (beta Cygni) - R.A. 19h 29' / DEC + 27 51 - Mags: 3.0 & 5.1 (wow!!)

OBJECT 3:   Very good double - Delta Cygni - R.A. 19h 43' / DEC + 45 00 - Magnitudes: 3 & 6.5 (great for small scopes)

OBJECT 4: Nice close double - Mu Cygni - R.A. 21h 42' / 28 31 - Magnitudes 4.7 & 6.1 - Good test for 4-inch!

OBJECT 5:   Orange Doubles! - 61 Cygni - R.A. 21h 04' / DEC 38 28 - Magnitudes: 5.3 & 5.9 very easy, interesting!!

OBJECT 6: QUAD star - Nu Scorpii - R.A.16h 09' / DEC (-)19 21 - four stars, two close pairs.

OBJECT 7:   My favorite variable! - SS Cygni - R.A. 21h 41' / DEC + 43 21 - Magnitude range - 8 to 11.7, 51 days!

OBJECT 8: nice galactic star cluster - Messier 29 (ngc6913) - R.A. 20h 22' / DEC + 38 21 - Mag: 7.1 small & nice

OBJECT 9: another galactic cluster - Messier 39 - (ngc7092) - R.A. 21h 30' / DEC + 48 13 - Mag: 5.2, 30 stars

OBJECT 10: *DUMBBELL NEBULA* - Messier 27 (ngc6853) - R.A. 19h 57' / DEC + 22 35 - (in Vulpecula!)

OBJECT 11: Interesting diffuse nebulae - ngc6992 & 6960 - R.A. 20h 54' / DEC + 31 30 - the *"Veil Nebula*!!"

OBJECT 12: galactic star cluster - ngc6866 - R.A. 20h 02 / DEC + 43 51 - about 50 stars, nice cluster!

OBJECT 13: *NORTH AMERICAN NEBULA* - ngc7000 - R.A. 20H 57' / DEC + 44 08 - NEED DARK SKY, no scope!

OBJECT 14: Very nice double star - Gamma Delphini - R.A. 20h 44' / DEC + 15 57 - Magnitudes: 5.5 & 4.5

A VISUAL GUIDE TO OUR DEEP SKY OBJECTS IN CYGNUS, Delphinius, Sagitta and Vulpecula –

<u>Object 1</u> - Very Bright Star Deneb (Alpha Cygni)
Let's start our "GO TO" TOUR of the Cygnus region with a slew to "Object/Stars/Named.... DENEB" This is a very deceptive star, appearing of average brightness and shining innocently against the soft glow of the summer Milky Way. But looks can fool. Deneb, at a distance of 1600 light years, is one of the largest SUPERGIANT stars we know of, only equaled by Orion's Rigel which is closer. At the distance of our own sun, Deneb (visual magnitude 1.3 and only the 19th brightest naked eye star) would appear an incredible 60,000 TIMES BRIGHTER! Like Rigel, it is a very "early" star, meaning beginning its evolutionary trail on the Hertzsprung-Russell diagram of star development.
To me, this star has always appeared distinctly "yellow" but it is truly a bright white supergiant just

like brilliant white Rigel. Just to the east of Deneb is the famous "*North American Nebula*," a very beautiful reflection nebula at about the same distance from us as Deneb is.....it is thought that the "glow" from this nebula (ngc7000 - see following) is actually a result of energizing from Deneb itself!

Object 2 - the BEST double star of the sky...ALBIREO (beta Cygni) - a wonderful contrast in color!
Although this is one of the most magnificent stellar sites in the cosmos, the name actually means...."nothing." It is an error it appears. Somewhere in the 16th century translations of all the earliest Arabic star and constellation designations, the name originated for this star as "Ab rireo" which really translates to nothing in particular. The actual recorded Arabic gives the star the name of the "Hen's Beak," or "Al minhar al Dajajah." Perhaps that was too difficult with too many "a's".

Albireo, at magnitude 3.1, is the most remarkable of all double stars in color contrast. The brightest star, magnitude 3.1, is clearly a brilliant yellow color, while the smaller (fainter at magnitude 5.1) is a distinct aqua or blue-green color. There is no mistaking the rich color contrast here, even if you are color blind. However, I must warn you that - even thought the star can be easily resolved into the two striking components in LOWEST magnification of a small telescope- you must use about 15x per inch aperture to actually appreciate the saturation of these deep colors. If you get too high a magnification, the color diminishes with the brightness....too low and the effect is gone from the intensity of the stars themselves. Albireo, at a distance of only 410 light years (the exact same distance as the Pleiades of late Autumn!), is much closer than Deneb is to Earth.

Object 3 - Delta Cygni - A Real Test Double for the 3-inch

This star has a primary component of a bright magnitude 2.8 and a secondary star that is only about 2" arc from it at magnitude 6.5, about SSE. At present, this double star is thought to be at maximum separation as seen from Earth, but still presents a challenging object for the 3-inch scope. It is likely unresolveable in anything smaller. Because of the difference in brightness, the 2nd magnitude star overshines your ability to clearly differentiate the fainter star, so use medium high (about 150x and slightly above) to begin to resolve this star.

Object 4 - Mu Cygni - A good double star....perhaps a severe test for the small refractor

This star only ten years ago was a much tougher double to split. However, its apparent separation has been increasing since it was first discovered and is now about its maximum, some 3.2" arc, making it an ideal test object for high power with the small scopes; the 4-inch and larger scopes should have no trouble seeing both components, magnitudes 4.7 and 6.1; look for the second star almost due NW from the brighter star. Both stars are "solar-type" in that they are much like our sun in luminosity and composition and only some 65 light years away. In the 6-inch and larger scopes, look for a very faint (11th magnitude) OPTICAL companion to these two stars about 48" arc away, about the size Jupiter would extend in the same medium-high power

312

eyepiece.

OBJECT 5 - Nice very wide Double Star, has actually MOVED across the sky since 1792!
This is a quite famous and very nice wide-field double star. In 1792 the Italian astronomer *Piazzi* noted that this already-known and cataloged double star was MOVING relative to the more distant stars behind it! Yes...it actually moves more than 5" arc each year toward the northeast! So in only about 200 years, this star has moved the equivalent distance of one-half the diameter of the moon when comparing to stars around it. The obvious reason for this is that the star is quite close to Earth, only about 10 light years away. It is a fabulous double star, with the pairs oriented in a true NE-SW direction from one-another and nearly equal brightness (5.3 and 5.9). Requiring an incredible SEVEN CENTURIES to complete one full orbit, the stars now appear a wide 30" arc apart, clearly wide enough to be seen comfortably at lower powers in all of our telescopes. Just for the record, in addition to the two stars we can SEE, 61 Cygni is being orbited by an "unseen" dark companion that possibly might be a proto-planetary mass of some type, about 8 times more massive than our own Jupiter.

Object 6 - **Dr. Clay's Favorite** Variable Star: SS Cygni
Here is an exciting, fun and unpredictable variable star that is somewhat "predictable." Looking at the light curve below, you can see that the star SS Cygni has an average "period" (from brightest, down to dimmest and then back to brightest peak again) of only about 51 days. It will routinely vary

from a bright 7.8 (or close) to almost magnitude 12.0, but the exact pattern is never certain.

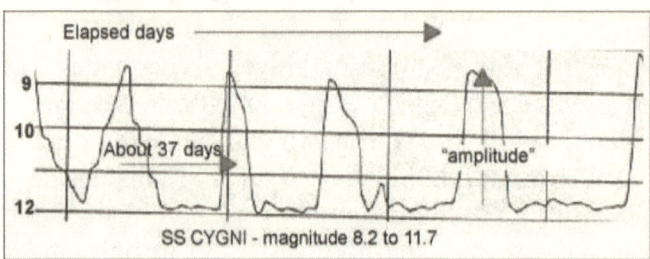

Even in this example for about four cycles, you can clearly see major differences in each period. SS Cygni is of the "dwarf novae" type (R Geminorum class) which is characterized by minor and predictable actual eruptions of the stellar "surface." Normally this star rests at magnitudes of about 11.8 to 12.0, but at intervals which can range from two dozen to nearly 100 days, it will suddenly and briefly erupt to its full brightness. Some maxima will last for a full 18 or 20 days, while others only about 8-10 days....but you never know which to expect. That is why this star is so interesting.

Because of its short cycle and the brightness range is in reach of all of the telescopes (although during its faintest part of the cycle it might be beyond the reach of smaller scopes) and the ease of which it can be located with the American Association of Variable Star Observers' (AAVSO) "reversed" chart which duplicates the mirror-image field of Maksutov and Schmidt-Cassegrain telescopes, this is an inviting and rewarding object for your efforts. Download the following chart from the AAVSO, save to file, bring up and resize to your page and print a full-sized star finder chart complete with verified comparison star magnitudes to begin YOUR research on this exciting star: https://www.aavso.org/apps/vsp/ . Note for these charts, simply type in the NAME of the variable at top to generate your choice of chart.

These chart sare the "a" and "g" charts, or moderately wide field and low power view. If you are new at locating and estimating variable stars, you might wish to download the "a" scale (wider field) chart from the AAVSO that will assist you in finding the star. Also note that, when SS Cygni is faintest, it might be good to move up to the "g" chart which closes in on the star and its proximity, showing fainter comparison stars and a more narrow field of view. Both of these charts are available on-line through the same address. Proper methods of observations can be found, along with a full discussion of variables from my guide: "*Observing Variable Stars*" in the GUIDES/ Observational tabs on the ASO website.

It is now know that SS Cygni is a binary star, as probably all U Geminorum "dwarf novae" star are likely. One component is a yellow dwarf star being encircled by a hot blue star, this thought to be the responsible star for the sudden outbursts as fresh gases from the yellow star stream gravitationally onto the hot surface of the intense blue star.

Object 7 - Cygnus X-1
In our "GO TO" tour of the constellation Scorpius, we look at our first "unseen" object, the highly intense X-ray emission of Scorpius X-1, a very faint 13th magnitude object just at the limits of the 8-inch. Here we have one that you can actually "see,"

315

in *Cygnus X-1*. Probably a black hole - and listed as one on your computer library of objects, *Cygnus X-1* shows optically as a rather uninteresting 9th magnitude star. It is only one-half degree ENE of the bright star Eta Cygni in the center of the swan. Recent studies show that the strong emissions from this star are NOT due to supernovae activity, but rather from a suspected "collapsed" object such as the black hole phenomenon. You can use the following photograph, courtesy Lowell Observatory (south at top), to locate this curious object. It MAY be the only "black hole" that you "never see."

FIELD OF CYGNUS X-1
Suspected Black Hole

Object 8 - Very Nice Galactic Cluster in the

Summer Milky Way - Messier 29

Only about two degrees SSE from Gamma Cygni, this nice galactic cluster, Messier 29 (ngc6913) is only one of dozens of such objects throughout Cygnus and this region of the Milky Way. Your first impression in a small low power scope will be a "rectangle" of equal brightness stars, surrounded by others scattered about the quadrilateral. There are about 20 stars in the 8th and 9th magnitude range, all visible in these scopes on a dark night. At about 1/3 the angular diameter of the moon, this is an easy object, although in larger scopes and at higher magnifications the "cluster" impression soon diminishes. So use the very lowest power and the widest field possible. This area of the Milky Way really deserves some scanning time, so deviate a bit from your busy GO TO schedule and stay a while. Messier 29 has always reminded me of a "little Pleiades", much smaller due to its great distance of 7200 light years (almost 200 times more distant that M-45!).

Object 9 - Another nice galactic cluster! - Messier 39

Messier 39 (ngc7092) is considerably closer (800 light years) than M-29, above and appears larger because of it. It was recorded as early as 320 B.C. by the Greek philosopher **Aristotle**, appearing as a glowing light in the sky, barely visible to his naked eye. It appears to me without any optical aid much like Messier 35, the large cluster in Gemini, except no so large. About 35 stars can be seen in this cluster, spread over a field about the size of the moon's diameter; hence, use ONLY low power and wide fields to observe. Stars can be resolved into component images in all telescopes, but the best image is likely afforded by a small scope at about 50x. About twice as far as the close Pleiades, this cluster is probably about twice as old as well, still making it a very young group of entries into the stellar population.

Object 10 - THE FAMOUS *"DUMBBELL NEBULA"* in Vulpecula
To me, this object is still in the constellation of Cygnus, just southeast of the beautiful double Albireo and only 3 degrees due north of gamma Sagittae. Although it may not be as popular as another planetary nebula, the famous "Ring Nebula" in Lyra (see my constellation Guide "GO TO.... *Lyra*" under the GUIDES tab here on this ASO website for complete details and drawings), this is the brightest and most conspicuous planetary nebula in the sky. It is very large (8' long by 5' wide) and appears to me shaped like a man's "bowtie" more than a "dumbbell." (Photo by ASO)

This photograph under a very dark sky shows M27 as photographed at ASO with a CCD camera, 30-seconds. This is far more than can be seen with just the eye.

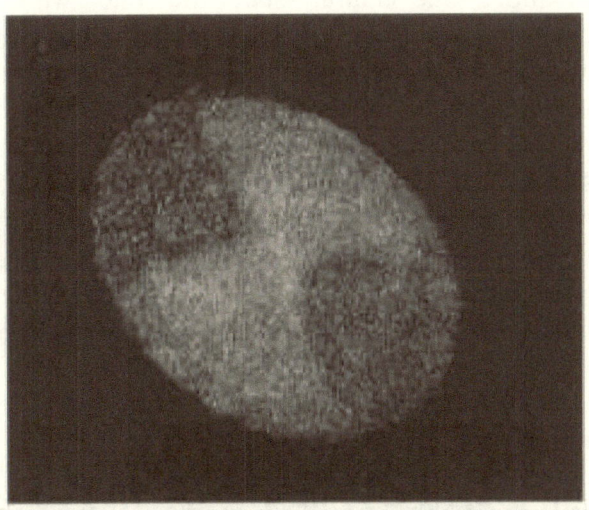

Drawing of the Dumbbell Nebula
by Lord Rosse - 72" Reflector
ARCHIVES OF THE ARKANSAS SKY OBSERVATORY
PLATE PUBLISHED IN 1849

Compare this to the mid-19th century engraving by **Lord Rosse** from his observations with the giant 72" reflector that was the giant of the world at the time. Because of the excitement over the newly-resolved globular clusters and some larger galaxy clusters, Lord Rosse apparently (and incorrectly of course) believe that the nebula was comprised of thousands of pin-point faint star images as can be seen dotted across his drawing. This drawing and MANY other similar engravings can be found in the classic book (1849) "*Letters on Astronomy Addressed to a Lady in which the Elements of the Science are Familiarly Explained in Connection* *with its Literary History*." WOW! You think the title is long....wait 'til you read the BOOK!

At about 850 light years, the large planetary nebula still exhibits its central star, the one whose explosion triggered the large shell of luminous gas that we see today. At magnitude 13.4, this star is almost equal to that "seen" in the Ring Nebula, although I have routinely seen the central star of Messier 27 in an 8" aperture telescope with difficulty (very easy in the 24"), but the star within Messier 57 is extremely difficult. This is curious since the gas cloud in MORE concentrated in the Dumbbell than in the "hole" of the Ring Nebula.

In the small refractors use about 90x minimum to observe Messier 27 so that its distinctive shape can be appreciated. The star field surrounding it at that power is incredible as well. In the 3-inch, the best views are provided at around 90x to 120x. My most satisfying glimpse with larger telescopes has always been with pretty hefty power - 227x.

Objects 11 - Two Gossamer Arcs of Nebulous Strings: ngc6992 and ngc6960, the "Veil Nebula"
This is a remarkable object that shows up very well on piggyback photographs of wide field constellation shots of Cygnus. These two diffuse gas clouds (actually one objects, the remnant of a ancient supernova explosion) are located in the far southeast corner of Cygnus, right at the border with Vulpecula, and only a few degrees north of the planetary Messier 27 (above). Although it took an 18" telescope 200 years ago to find this object, it can be seen in modern scopes today; if you have very dark skies, no moon and the 6-inch and larger

scopes, you can spot this arc of nebulosity in its true shape. LOW POWER is a must, as higher magnification increases sky contrast to mask out the faint clouds of gas. NGC 6992 is the brighter of the two and will appear on the "upper right" (northeast portion) of the field of a wide angle view; the dimmer object, NGC 6960 is a real test for a 5-inch but can be held steady in the 8-inch in dark skies. Both appear to be semi-circular arcs of very faint light that "almost" intersect with one another visually. In all this object is a ring of gas debris almost 3 degrees in diameter! It is believed to be the shell of yet another supernova blast from a now-annihilated star some 150,000 years ago.

Object 12 - Very nice, small glittering of stars: Galactic Cluster ngc6866

Although not an easy object, this very small (only 6' arc) galactic cluster of some 50 stars is located nearly due west of the "heart" of the northern cross. In the smaller scopes expect only some 10 stars to be clearly distinguished, with the remaining appearing as an interesting glow among an absolutely breath-taking star field. Use about 25x per inch for best views of the cluster in all scopes, but much less magnification and wider field to scan the star field around it. In the 6-inch about 20 stars can be seen clearly and about 10 ore in the 8" aperture. This is a very tiny object, but one worth looking for. At first, at very low magnification (which is the best way to locate it) it will appear only as a faint glow like a little comet without a tail; only with some "power" will you begin to make out individual stars.

<u>Object 13</u> - The *NORTH AMERICAN NEBULA*! - ngc 7000

Somehow, perhaps as a fitting tribute to Uncle Sam, this remarkable diffuse reflection nebula was one of the "lucky NGC objects to get the "*millennium index*," the even-numbered designation of "7000." After seeing wonderful photographs of this objects, most amateurs expect to be able to see it. After all: it is easy to find (only 3 degrees east of bright Deneb!); it is **HUGE**...120' arc x 100' arc!; and it is bright - 1.33 magnitude. Ahhh....but the downside to all this is, indeed its SIZE. It is so large that the brightness is spread over a surface area more than *90 times that of the entire moon*! So you are lucky to see it once in a lifetime.

NGC 7000
North American Nebula / Pelican Nebula
near Deneb in Cygnus
200mm F/2.8, 40 minutes - SO 155 Kodak

This object is one in which a small qualilty fast APO refractor can beat the rest of them hands down. Using incredibly low power and ONLY on the darkest night atop some remote mountain....no moon....this object will stand out in an almost 3-D appearance against the brilliant star dust skies of the Milky Way; it can be seen in a very good pair of 10 x 50 binoculars under idea conditions. Wide field, piggyback photography (see the ASO GUIDES/ General for my introduction in to piggyback photography and how easy it is with your telescope!) reveals the entire nebula, including the "Pelican nebula" (to the lower right in the above photo....the "pelican" facing Mexico!) and even the dark "Gulf of Mexico!" complete with the Yucatan Peninsula! DO NOT EXPECT to see this faint glowing gas cloud in a large high-powered telescope.

Object 14 - A Fine Double Star in Delphinus
Gamma Delphini
At magnitude 3.9 this is perhaps the most interesting object (and a good one!) in our celestial Porpoise. Gamma Delphini makes up the northeast "corner" of the small diamond comprising the "body" of the dolphin. With a nice wide separation of 10" arc and two fairly bright stars (magnitudes 4.3 and 5.1) this is an outstanding object for all telescopes, but particularly nice in the smaller, low power scopes. I find the brighter star to be somewhat "orange" color and contrasting nicely with the dull white or even "gray" color of the fainter star. It truly is a very nice double for contrasting and interesting color combination. For these telescopes, it is recommended to use about

80x to 100x for best views and renditions of color. For the 6-inch and larger scopes, there is a challenging double star just 15' arc (half the moon's diameter) due south of this pair. At magnitudes 7.5 and 8.0 this evenly matched double is about 5.7" arc separation, so it CAN be resolved with the 4-inch as well in good skies. Best resolution will occur at about 25x per inch aperture on this faint star, located at R.A. 20h 44' / DEC. +15 degrees 43'; it can be spotted in the very same field with Gamma Delphini in low power.

WANDERING ABOUT....YOUR NEW "USER OBJECT" IN DELPHINUS

THAT'S RIGHT....not Cygnus. Delphinus. And there is a reason for it. If we do NOT put a User Object into your computer telescope library in Delphinus....you may NEVER look at anything in that constellation again! It's a DOLPHIN for crying out loud! Show your support for our intelligent watery counterparts and their fragile existence in the environment! GO TO a double star on the fish's snout!

On AutoStar or PC handbox (use the proper tabs/keys for your exact sky program), go to: "Select/Object [enter]...." scroll down to "User Object" [enter]. Now enter the coordinates given above for "Gamma Delphini", using the number keys on AutoStar. After entering and pressing "Enter" yet again, scroll down one and you can list the magnitudes of both stars [Enter].

Please remember that the exact process and keystrokes will vary from program to program!

Chapter 13

DELPHINUS and EQUULEUS
An Odd Menagerie of Animals
The Leaping Celestial Dolphin Rides Herd on the Little Sky Horse

Our next Constellation Guide, "GO TO DELPHINUS & EQUULEUS" takes us to two very small, but not often overlooked constellations. Although Equuleus (the "Little Horse") is either not known about entirely to casual stargazers, or is frequently ignored (you will see why), its partner in the celestial zoo, DELPHINUS (the "Dolphin") is very conspicuous and forms an interesting **"asterism"** (star pattern) in the summertime skies.

Although these two constellations are seemingly void of any bright "show-stopper" deep sky objects - and so far from the ecliptic that they are never visited by any of the major planets nor moon - there ARE many interesting doubles stars and some very fine long period variable stars that are worth checking out. For a complete list of the double stars, I highly recommend the Burnham's Celestial Handbook, Vol.. Two which lists nearly all observable double stars, coordinates, magnitudes, separations and many times good descriptive notes in "shorthand" form. For the list of variable stars (Burnham's also has a listing of these) as well as FREE observing and comparing charts, I recommend www.aavso.org , the **American Association of Variable Star Observers** in Massachusetts which has supported visual (and photoelectric) observations of these stars for nearly a century. There is much free (or very inexpensive)

Material available from that organization.

*   *   *

Clay Sherrod's
CONSTELLATION GUIDES:
"Go. To" DELPHINIUS & EQUULEUS
The "Dolphin" and The. "Horse"

VULPECULA

SAGITTA

PEGASUS

DELPHINIUS

10
7006

4

2 Svalocin

12
6891

1
Rotenev

9

AQUILA

EQUULEUS

6

11
6934

7

3
Kitalphar

Bulletins of the

ARKANSAS

Lat.
+35 05

Sky

Lon.
-92 30

OBSERVATORY

AQUARIUS

A finder chart for locating many of the GO TO
objects in the constellation Delphinus and region; if
using a computer planetarium program, you are
encouraged to plot the objects on your screen for
higher resolution than this chart provides.

*   *   *

Equuleus (pronounced "u-QUE-lee-us") looks nothing like a horse....perhaps the early Greek philosophers had a horse left over that they needed to pay tribute to and found this random group of stars "left" over from all the other nearby constellations and decided that would be it. On the other hand, Delphinus appears strikingly like a leaping dolphin in the night sky!, its arcing tail formed by the brighter stars Eta, epsilon and kappa, while the body and head of the dolphin are formed by alpha, beta, gamma, delta and zeta (see the second chart "*Job's Coffin*" below).

Notice the ONLY two star names in the constellation of Delphinus. Do they appear to be somewhat "different" than those of previous constellations? Are they Russian? Slovakian? Nope....they are BACKWARDS! Most of the bright stars discussed in these "GO TO" TOURS were named centuries ago by ancient Arabian sky watchers who provided us with beautiful, colorful and meaningful star names. Although the exact reason that this happened for these two stars is not know (there is some pretty wild conjecture, though), the names are actually BOTH first and last names of **NICHOLAS VENATOR**, a night assistant at a major observatory (Palerma Observatory) in the first decade of the 19th century. Whether he did the trickery, affixing his own backwards-spelled names to these two previously unnamed stars when the he Palermo Star Catalog was published in 1814....or whether someone else did it for him is not even important anymore. These stars will now ALWAYS be known as "*Sualocin*" and "*Rotanev*."

The Dolphin was very important to the preservation of Greek culture, as the legend states that Delphinus (pronounced "del-FIN--us") rescued and whisked away the imperiled Greek poet *Arion* from his enemies who were chasing him down for mortal harm. I am reminded of a fabulous Greek restaurant named "*Boy on a Dolphin*" in Pensacola Beach, Florida that was so-named as a result of this wonderful legend. This story was widespread for whatever reason, and it appears throughout the lore of Sicily, Southern Italy and throughout Athenian lore.

Clay Sherrod's
STARS OF DELPHINIUS
"Job's Coffin"
BRIGHTNESS OF THE PRINCIPAL STARS
OF THE CONSTELLATION

Decimal Points are Omitted to Differentiate from Stars (i.e., "377 = 3.77"

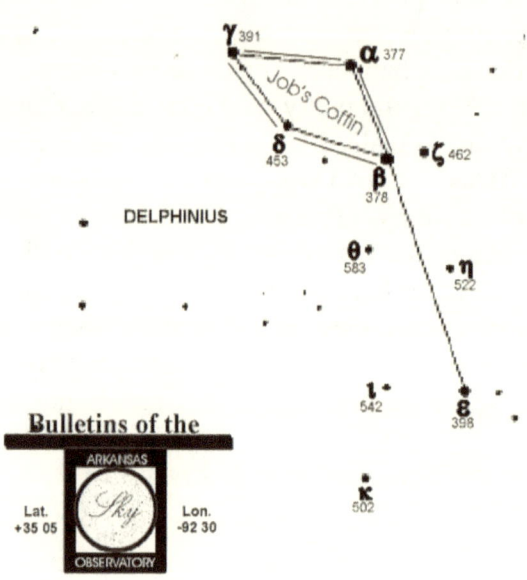

Very interestingly, even the early Indian and Hindu cultures recognized this small constellation as a *"Porpoise"* (also a dolphin), but the Arabians, who missed a bet on naming those two stars and saving us a lot of grinning, knew the constellation as the *"Camel who rides"* (can be ridden).

In Hebrew writings the constellation and its stars, like so much of the night sky, has connections to the writings of the Old Testament, with this star group denoting the Whale that devoured unfortunate Jonah. After the death of Christ, early Christian groups envisioned the *cross of Christ* in the small "body" of stars that makes a diamond-shape.

More recently, and amusingly, this same diamond pattern (see the chart immediately above) today is often referred to as *"Job's coffin,"* and it indeed, does look like the old pine box of the wild west and early European cultures.

The stars of Delphinus are ideal for making visual magnitude estimates of bright variable stars nearby in summer skies. Also, use these magnitudes I have provided above for the valuable determination of METEOR magnitudes during the many meteor showers of summer and early fall. Merely compare the brightness of the "shooting star" to that of the closest star given in the chart and writing down (I use simple hash marks) each meteor for each given brightness group. The total number at each magnitude is important to understanding the nature and density of many of the meteoric clouds and their tiny fragments that result in the fine meteor displays.

Note from the sky chart included here that the *CELESTIAL EQUATOR* passes way SOUTH of both Delphinus and Equuleus, and thus all angular measures (declinations) are NORTH of the celestial equator and positive ("+"); hence you will see references in this "GO TO" GUIDE to "(+)" declinations for all celestial objects in both constellations.

As with every "GO TO" tour guide, each GO TO object in this area is discussed for your telescope regarding the type of conditions necessary for you to view it optimally for discern the very faintest details.........magnifications and aperture necessary for most objects, and much, much more. This is YOUR complete guide to get you on your way to exploring the best (and few!) objects in these two constellations. The following listing of "BEST" objects contains the finest or most interesting from my own observing experience and preference.

Use the attached star chart and the following Guide as an excellent reference for your next star party itinerary, or a beginning for further study into the thousands of objects visible in this part of the sky. Truly these extensive Constellation Study Guides will most definitely put your computer sky library to work for you in the most efficient and enjoyable way possible! As a matter of fact, MANY AutoStar or PC sky program users are now programming their own "Tours" based on these guides, using each constellation as a separate GO TO Tour for the AutoStar or sky program library that can be added in or deleted through the main edit screen on your PC or MAC computer.

I hope you enjoy this comprehensive GUIDE to touring this constellation via your PC sky program and its computer-driven telescope. Each new installment is complete with diagrams, charts and illustrations that you will find nowhere else. Please let us hear YOUR feedback and your observations of each and every constellation after YOU have toured its vast reaches of our skies!

YOUR DELPHINUS AND EQUULEUS CONCISE DIRECTORY OF INTERESTING OBJECTS –

Although there are slim pickings to choose from in both of these constellations, I have chosen the finest (or most interesting) 12 objects in this DELPHINUS / EQUULEUS "GO TO" TOUR; as with all GUIDES, all objects listed below will be visible in most telescopes (some naked eye) from a 3-inch through 8-inch; of course larger apertures may "show" an object a bit closer and "better," but frequently a wide field and low power view is more desirable than aperture for FINDING the objects initially. And remember that not ALL objects listed are visible in some telescopes.

If you have difficulty identifying any object I strongly encourage you first FIND the target object, or its approximate location through your GO TO function with your lowest power and then - once IDENTIFIED positively - move up slowly in steps with magnification if necessary. Remember, not all objects "like" magnification. If, after you have reached a beautiful view of your object, more magnification results in 1) dimming of the object; 2) loss of contrast; 3) loss of color, then you have exceeded the optimum magnification for that object.

The rule for determining "optimum magnification" is that: 1) too low power results in sky background glow detracting or diminishing the contrast against the deep sky object; 2) too high magnification darkens BOTH the sky background AND the object; 3) medium magnification can be achieved at which you have MAXIMUM contrast between the object and its darkened background sky. I have found through three decades of direct observing that about 15x per inch aperture for deep sky observing is PERFECT for most objects. That being said, always remember that DOUBLE or multiple stars require whatever power you can crank out....the seeing conditions are the limiting factor here.

For my complete and comprehensive discussion regarding seeing conditions and sky transparency, see my GUIDE at the tab: GUIDES/General on the Arkansas Sky Observatories web page.

As with all of the "GO TO" tour constellation lists, I recommend a good star atlas and/or chart which will list all the finest objects, constellation-by-constellation. One very handy reference guide is the *PETERSON FIELD GUIDE TO THE STARS AND PLANETS*, which features complete lists with declinations, right ascensions, magnitudes, and all pertinent information for you to expand your observing horizons beyond this brief guide. For the computer or media device, my favorite quick reference and very comprehensive descriptions of objects are the many Smart Phone APPS that are available such as Voyager, etc.. Note that this very affordable software package ALSO allows very quick and accurate construction of "Go To" Autostar tours for these constellations as well as allows a very nice laptop control of any telescope via Bluetooth or other wireless connection!

The constellation tour Star Chart, preceding, will get you started on your journey for this constellation.

With all deep sky objects, avoid attempting to observe when the moon is in the sky, even a very thin crescent, as its brightness in the sky will overshadow the very dim contrast afforded by even the brightest deep sky object; if you see the object at all against moonlight, you will NOT see the subtle outlying areas or the full detail of what is presented.

The high declinations of Delphinus and Equuleus offer good and long-period observing for observers users in the northern hemisphere. All deep sky objects and difficult double stars are ALWAYS best observed when they are located nearly overhead (or as high in the sky as possible), thus requiring the observer to look through the thinnest portion of the Earth's "lens" of atmosphere and haze.

Following is the concise object list for your "GO TO" TOUR of DELPHINUS and EQUULEUS; you may wish to find the majority of the objects from the sky program Library (for example, you can easily go to NGC 7006 if you pull up "Object/Deep Sky/NGC/..then type in '7006'...." and then press "Enter", followed by "GO TO" to access this FAINT but very enjoyable globular cluster On the other hand, if you want to experiment and become a "better computer user" try entering the exact R.A. and DEC coordinates of that object as described above after holding down the MODE key. You will find the accuracy of entered GO TO's to be somewhat less than those stored in you PC program, but the capability of acquiring unlisted objects is fantastic!

OBJECT 1: brighter star - ROTANEV (beta Delphini) - R.A. 20h 35' / DEC + 14 25 - Magnitude: 3.8
OBJECT 2: brighter star - SUALOCIN (alpha Delphini) - R.A. 20h 37' / DEC + 15 44 - Magnitude: 3.8
OBJECT 3: (Equuleus) brighter star - KITALPHAR (alpha Equulei) - R.A. 21h 13' / DEC + 05 02 - Magnitude: 4.1
OBJECT 4: (Delphinus) good double star! - gamma Delphini -R.A. 20h 44' / DEC + 15 57 - Magnitudes 4.5 & 5 - yellow stars
OBJECT 5: (Delphinus) nice variable star - V Delphini - R.A. 20h 46' / DEC +19 09 - Mag. 8.2 to 13.6 - good for 6-inch scope+
OBJECT 6: (Delphinus): interesting nova - HR Delphini (nova Del 1967) - R.A. 20h 40' / DEC + 18 59 - Very unusual and active star
OBJECT 7: (Equuleus) very nice double - 2 Equulei - R.A. 21h 00' / DEC +06 59 - Magnitudes: 7 & 7 - great star in 3-inch +
OBJECT 8: (Equuleus): good quad star! - 5 Equulei - R.A. 21h 08' / DEC + 09 57- Magnitudes: 4, 11, 12, 6 - interesting star
OBJECT 9: (Equuleus) variable star - R Equulei - R.A. 21h 11' / DEC + 12 36 - Magnitude: 8.7 to 14.8, 261 days

OBJECT 10: (Delphinus) globular cluster - ngc7006 - R.A. 20h 59' / DEC +16 00 - Magnitude: 10.7, most remote of all globulars!

OBJECT 11: (Delphinus) globular cluster - ngc6934 - R.A. 20h 32' / DEC + 07 14 - Mag. 9, very small but easier than ngc7006

OBJECT 12: (Delphinus) planetary nebula - ngc6891 - R.A. 20h 13' / DEC + 12 35 - Mag. 10 with 11th mag. central star! 3-inch scope+

A VISUAL GUIDE TO OUR DEEP SKY OBJECTS IN DELPHINUS AND EQUULEUS

Object 1 - Our "Starting" Brighter Star –
"ROTANEV" (beta Delphini)
Our starting point for every "GO TO" TOUR is always (or usually!) the brightest star of the constellation or region but in Delphinius, the TWO brightest stars share nearly the same magnitude. I have already relayed the story (true) on how these two stars were assigned their names, albeit perhaps illicitly by scientific standards. *Rotanev* (beta Del) is magnitude 3.7, while *Svalocin* (alpha Del) shines at magnitude This is by far the DIMMEST bright star we have used as our "starting target" for any constellation "GO TO" tour, surpassing the "dim bulb award" of *Zuben El Genubi* from the inconspicuous Libra constellation. *Rotanev* is actually a pair of 4.0 and 4.9 stars, orbiting VERY close to one-another; a keen-eye observer with the 8" scope or larger MAY be able to resolve these two, at only about 0.7" separation.

Object 2 - The other "Brighter Star" in Delphinus:
"SVALOCIN" (alpha Delphini)

This star and Rotanev - as well as gamma and delta Delphini - "may" be part of an actual star "association," a configuration of stars all at the same distance that appear to be moving synchronously or gravitationally bound to one-another and hence a very "loose and sparse" cluster of sorts. All are moving at about 12 kilometers per second TOWARD earth, and are presently some 125 light years distant.

Object 3 - Yet another "Brighter Star" –
"KITALPHAR" (alpha Equulei)
Of course ALL stars in the sky should be interesting and important to us....it's just that this one is really not. I have placed this star on your "GO TO" TOUR merely to allow you to familiarize yourself with the otherwise easy-to-miss constellation of Equuleus, and because frankly....there is just not very much in that constellation to look at. Even if you spell THIS star backwards ("RAHPLATIK") it still means absolutely nothing (other than perhaps a good name for a rock band). At least NOW you know where Equuleus is! AT magnitude 4.14, it is actually a spectroscopic double star, with both components very much like our sun on the evolutionary scale. This star is VERY similarly placed (45 parsecs) as are the four stars of "Job's Coffin" in Delphinus and may well be part of this loose association.

Object 4 - A Very Nice Double Star – Gamma Delphini - Nice Colors!
Finally, amidst all the seemingly uninteresting primary stars of these two constellations is something of better-than-average interest! Gamma Delphini is an excellent double star, making the northeast corner of "Job's Coffin." This is a true

binary star (both stars orbit one-another), with the primary star (magnitude 4.3) being an easy 10.3" arc (about one-fourth the size of Jupiter's disk in the same eyepiece) nearly due NORTH of the 5.1 magnitude secondary star. This pair can be resolved with high magnification in the ETX 60 and 70 and with relatively low powers in the 3-inch and larger scopes. Look at the colors of these two stars; most observers see the brighter star as a pure "yellow" color, while the fainter star seems to have a green glow to it, a rather "greenish-yellow" color.

Object 5 - A Good Long Period Variable Star - "V Delphini" - (beyond limiting magnitude when faintest!)

This is a GREAT star to PUSH the limiting magnitude extreme on the 6 and 8-inch scopes. The star, when at brightest every 534 days, can easily be spotted even in a 3-inch, followed about halfway through its very rapid plunge toward minimum, and then fading from even an 8-inch as it approaches minimum. Even though it disappears during minimum in all of our ranges, it still bears watching since - even though it is "classified" as a long period variable (see my discussion on variable stars at ASO GUIDES/Frequent/*Observing Variable Stars* - the star goes through VERY erratic changes in its light curve. For example (see diagram below), the star's peak may be delayed from one cycle to another by as much as 20-50 days; many times the star will attain a maximum brightness near 8.0, and sometimes will BARELY top 11th magnitude! You really never know what to expect out of V Delphini!

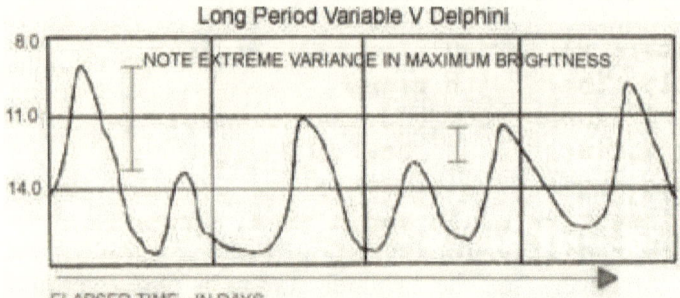

Long Period Variable V Delphini

NOTE EXTREME VARIANCE IN MAXIMUM BRIGHTNESS

ELAPSED TIME - IN DAYS

To both find and follow this star requires excellent star charts and comparison stars with magnitudes that have been carefully confirmed; such can be obtained from the American Association of Variable Star Observers through the link: https://www.aavso.org/apps/vsp/ . Note for these charts, simply type in the NAME of the variable at top to generate your choice of chart –

which will provide the "a", or low power wide field finder chart to first located the exact spot for this star; this chart is ALSO used to locate the NOVA "HR Delphini", described below! Once found, and when the star is near brightest, you will want to actually make your magnitude estimate comparisons by using the "a" chart at the same link. However, with the 6 and 8-inch scopes, the star can be followed via the "g" chart, which gives very faint stars and their magnitudes to monitor when the star is approaching minimum brightness. Note that there are several OTHER variable stars posted on the "a" chart; this should encourage you to monitor those stars as well! Full details about variable star reporting can be obtained from the AAVSO as described at the beginning of this "GO TO" tour. For V Delphini - since its period is just under TWO YEARS for a complete cycle -

estimating the brightness about every week or ten days is adequate.

Object 6 - *A Nova That Won't Go Away*!
HR Delphini - "nova Delphini 1967"
With the AAVSO "a" star chart
https://www.aavso.org/apps/vsp/
that you downloaded above, you can find HR Delphini, a nova that reached naked eye brightness in July 1967, rising from a star only magnitude 11.9 month before! It remained at this very bright maximum for nearly the remainder of 1967, with yet another outburst in November that put it at a bright magnitude 4.8, and another around Christmas that shot the star up to a brightness of 3.5! After two years, the star had only faded to magnitude 8.3 and even today it fluctuates around magnitude 11.2 to 11.7. Hence, it is still easily observable in a 4-inch scope using the "a" chart for that star. This is a remarkable star, a great example of a nova within our own galaxy and once which is STILL visible in amateur telescopes! If you do not monitor but ONE variable star during the summer months this one - and SS Cygni (see my Cygnus "GO TO" TOUR in the GUIDES/Constellation tabs on this website - should be the star. You will not be disappointed, with minor light changes you can easily record with every observing session!

Object 7 - A Double Star in a Little Horse - "2" Equulei
This is a very nice pair of stars, both 7th magnitude and maybe a bit hard to find because of their "dimmer-than-average brightness of most doubles on these "GO TO" tours of constellations. At equal brightness, look for this relatively easy (with 3-inch

and larger scopes) double oriented in about a north-south position relative to one-another; both are type "F" stars, similar to our sun in many respects. The current separation is about 2.8" arc which might make it a challenge for the 3-inch at 7th magnitude; however, the 6-8 inch should have NO trouble resolving this very nice pair, provided that your GO TO was pretty accurate; remember the rule: start off with low power and center what you "think" is the target; then gradually increase your magnification until you can verify the star is double.....this may take about 150x with the 3-inch and perhaps the same with the larger scopes, although the keen-eyed should see both stars at about 70x. This is not a good object for smaller telescopes.

Object 8 - Another Good Multiple Star in Equuleus - "5" Equulei, a Quadruple Star
This is a not-so-easy Quad-star, or 4-star stet; the primary star "5 Equulei" is 4th magnitude and fairly easy to identify. Okay....now you have found the easy part. Look DUE WEST of this 4th magnitude star only 1.9 arc seconds (remember, the theoretical resolution limit of a 3-inch SHOULD be about 1.3", so you CAN resolve this star!) and try to spot an 11th magnitude (that's right - 11th) tiny image of the first companion star. THAT one is star #2. It will be VERY difficult, except in scopes larger than 8-inches, because of the brightness of the main star. High power helps, about 150x is ideal. If you have a cross hair eyepiece, block the brilliance of the 4th magnitude star with a cross hair while looking for the faint star; that will greatly help. NOW: look for star #3 - at 12th magnitude! - almost DUE NORTH of the 4th magnitude star, but this time a whopping 48" arc, the same distance that Jupiter is large in the

same eyepiece. This can only be seen under very dark night skies, and perhaps in the 3-inch.....it should be a relatively easy star to find in a 6-inch and larger scopes. Three down....one to go. STAR #4 is the easiest to spot (other than the main star), as it is 6th magnitude and VERY far away - 6 minutes (not seconds) arc! in a SSE direction. Star #4 and the primary star will make a good pair even in a small refractor and should be seen in medium power in all scopes. Whew! That makes me star-whipped just thinking about it!

Object 9 - A Great Variable Star - R Equuleus
Like V Delphini, this long period variable has a tremendous magnitude range, stretching from its brightest at magnitude 8.7 to dimmest at 15.0. Thus, at dimmest, do not expect to see this star in any telescope less than 16" in diameter. Nonetheless, it can be tracked with a 4-inch through about 20 percent of its 261-day cycle, about 45 percent of it in the 6-inch and some 70 percent in the 8-inch. For a finder chart, log onto https://www.aavso.org/apps/vsp/ . Note for these charts, simply type in the NAME of the variable at top to generate your choice of chart –
and select first the "a" which is the "finder" chart that I often speak of, the "R" denoting a "reversed" chart to correspond with the image orientation of catadioptic telescopes with NORTH at top and EAST at right; most of R Equulei's cycle will require the "g" chart, however, the "d" chart being a much fainter magnitude comparison star field.

Object 10 - Oh Boy! A Deep Sky Object! Distant Globular Cluster ngc7006

Yes, there ARE deep sky objects in Delphinius and Equuleus.....just not many of them and not very bright ones.

NGC 7006
The Most Distant Globular
Cluster Known?
Palomar 48" Schmidt Photograph

This is perhaps the most distant of all globular clusters - 185,000 light years away from Earth. It shares that distinction with another such cluster, ngc2419 in the constellation of Lynx. It is highly possible at that distance that these two clusters are "shared" in a common gravitational bond by both our Milky Way and the Large Magellenic Cloud, or at least that the clusters MAY NOT BE associated with either at all! They may truly be non-galactic objects, perhaps the only objects known to NOT be associated with a galactic system in some way. This globular will be mistaken for a star in a 3-inch because of its tiny size of only 1' arc!

You can appreciate how tiny this object is when you realize that the famous globular Messier 3 is a full 24' arc across, almost the size of the full moon's disk! Thus, ngc7006 is only 1/24th the size of Messier 13, as compared in the photo composite that I created above showing both to actual scale. Indeed, you can see there is QUITE a difference! Hence, do NOT expect to see any stars with any telescope of any size in ngc7006, other than the world's giants. The photograph above of ngc7006 was taken with the large ASO Astrograph and just barely shows a few perimeter stars. This 11th magnitude globular will NOT be seen in smaller scopes and likely will be missed in a 3-inch; the 6-inch will show a very tiny fuzzy "star" in its position as will an 8-inch, only a bit brighter.

But....if you DO see it, you have the distinction of saying you have seen perhaps the most remote globular cluster of the galaxy in which we live!

Object 11 - Another Globular Cluster in Delphinus - ngc6834
Although no Messier 13 by any means, this globular should comes as a relief to you after attempting to zoom in on ngc7006. ngc6834 is a bit larger - 3' arc across - a and brighter at magnitude 9.0. Thus, it will be unmistakable (although still star-like) in the 3-inch; a glowing "fuzz ball" can be expected with the increased light gathering and image scale of the 6-inch and larger scopes. This globular is by no means a showstopper, but owes its increased visibility to being only one-fourth the distance as ngc7006.

Object 12 - A Planetary Nebula! - ngc6891
This is an object for the 6-inch and larger scopes only, at magnitude 11.4 and spread out over a diameter that is the same size as Jupiter's disk (44" arc). By comparison, the famous Ring Nebula (Messier 57) boasts a magnitude of only 9.3 but a huge disk size of TWICE that of ngc6891. This will appear as a very faint glow of light and requires about 170x to 225x to view adequately; make sure you have located the object first in medium power (about 100x) and then gradually increase the magnification. This planetary nebula - a gaseous shell expelled from the explosion of an unstable star (magnitude 14.2 and still visible) - is nearly FOUR TIMES more distant than the Ring Nebula.

WANDERING ABOUT....YOUR NEW "USER OBJECT" IN THE DELPHINUS-EQUULEUS REGION

I doubt it is any surprise that our new "User Object" for the "GO TO" TOUR of Delphinus and Equuleus is NOT going to be a deep sky object! You have been exposed to all three deep sky objects of any interest to telescopes less than the Hubble, and there is not much left to tantalize you in these two "object-starved" constellations. But I DID intentionally leave out the BEST double star (actually a "quadruple star" for the 8-inch and maybe the 6-inch) for amateur telescopes in this area, EPSILON EQUULEUS at Right Ascension: 29h 57m, Declination + 04 degrees 06m. The primary star is Epsilon Equuleus, at magnitude 5.5, which should make this an easy target to begin on. STAR # 2 is the closest, and the one that will most definitely be NOT seen in the scopes smaller than 3". It is a bright star of magnitude 6.0 almost DUE WEST of the primary star by about 0.9" arc; this is most definitely at the 5-inch limit of resolution but I have seen this star on numerous occasions with powers in excess of 227X; my 7mm Plossl in a good 6-inch telescope shows clean separation between these two, although not much of it! It is not easy in the 8-inch either. STAR # 3 is a 7.0 magnitude star that is MUCH easier, and is likely to be seen in all telescopes at 11" arc away from Epsilon, nearly DUE EAST (and a tad north). This pair can be seen in a good small APO refractor. NOW.....STAR # 4 is a toughie and likely is NOT going to be seen except in scopes larger than 8", even though at magnitude 12.5 is "should" be visible in a 5" for observers with high altitude and very dark skies. Look for this elusive faint star 1.2' arc (minutes, not seconds.....so it is a good distance away, more than the disk of Jupiter at the same power) nearly DUE WEST (just like star # 2!, but just a bit further south). GOOD LUCK on this interesting foursome of stars, all part of this true multiple-orbit star system.

On AutoStar, or your PC program, go to: "Select/Object [enter]...." scroll down to "User Object" [enter]. Now enter the coordinates given above for "Epsilon Equuleus", using the number keys on AutoStar. After entering the coordinates and pressing "Enter" yet again, scroll down one and you can list the magnitude of the object as "5.5"[Enter].

Now you have at least ONE quadruple star on your AutoStar listing in its library! You might not be able to actually SEE all four stars, but it sure creates a nice "talking point" during your next star party....just verbally describe where each and every star ".....should be" to your visitor. It is sure to please!

*   *   *

"A wise man shall overrule his stars, and have a greater influence upon his own content than all the constellations and planets of the firmament...."
Jeremy Taylor

Chapter 14

DRACO
Ursa Minor and Cepheus
Dithering in the Domain of the Dangling Celestial Dragon...

In this combined constellation guide installment, "GO TO DRACO, Ursa Major and Cepheus"- of our constellation study guides for GO TO telescope users - we will explore a fabulous selection of fine nebulae, some unusual galaxies and some very nice and often overlook double and multiple star systems that are ideal for all telescopes. As will all of our "GO TO guides", this guide begins with a "start" of an easy GO TO to the bright star *THUBAN*, in the long and meandering constellation of DRACO and then proceeds through the many WONDERFUL interesting objects within reach of YOUR telescope, not only in Draco, but the surrounding constellations of **Ursa Minor** (yes, there is something to see there) and **Cepheus**!

This Guide encompasses Draco - a "dragon" or sometimes imagined as a "water dragon" - , Ursa Minor - the "lesser Bear" - and *Cepheus* - a very kingly "King" and husband of philanderer *Cassiopeia*. From bright, beautifully-colored and complex multiple star systems, to very challenging planetary nebulae, there are discussions along the way to tell you what to expect from each telescope size and type. All objects will be discussed with exact descriptions of what the viewers with any telescopes should expect to see...and what to NOT expect to see!

As in all Guides, useful magnifications for EACH GO TO object are discussed for certain telescopes, what type of night and conditions are needed to see certain details, double stars that can be resolved in each telescope model, and much, much more. It is your complete GUIDE for your deep sky observing pleasure and a very handy tool for use at your next star party!

Needless to say, it WILL put your GO TO telescope to work for you in a most efficient and enjoyable way!

I hope you will enjoy these comprehensive ASO guides to touring the *Constellations* which will features the most prominent constellations, complete with diagrams, charts and illustrations. Please let us hear from you with summations of YOUR observations through these constellation tours!

INTRODUCTION

The constellation of DRACO (pronounced "DRAY-co") is a large, winding and meandering string of stars that pretty much fills the high northern skies of spring and summer. The slithering dragon has been of much greater significance in history than we see it today, particularly in the mysticism of the Egyptian times of the Great Pyramids and Pharaohs.

The skies of **5,000 years ago** were noticeably different than they are today; yes, the "fixed stars for the most part were right where we know them to presently be, but the EARTH itself was tilted to a slightly different vantage point....an aspect of "*precession*" that is a result of the Earth's axis pivoting over a long period of time, like a wobbling toy top. As it does, the axes of the Earth - both north and south - will point to different parts of the sky.

That's right....*Polaris* has NOT always been our "north star." Boy Scouts of ancient Pharocian Egypt would have to learn another way of reckoning "north" than by the bright dipper stars *Duhbe* and *Merak.*

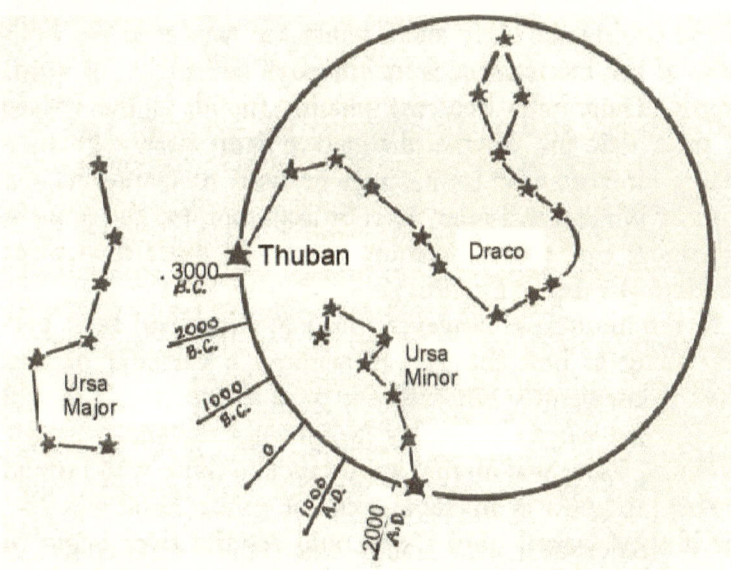

Showing the precession of the Earth's axis over time, making changes in which star becomes our "North Star"

Indeed, ancient sky watchers - the Priests - of Egypt about 3,000 B.C. were much aware that the not-so-bright star **THUBAN** was the "north star" of those skies long ago. The diagram above demonstrates the constellations of Draco, Ursa Major and Ursa Minor as they appear to "move" (actually it is the Earth that is wobbling and pointing differently, not the constellations) and hence rotating the honor of hosting the "Pole Star" in this past 5,000 years.

Interestingly, in "only" **15,000 MORE years**, this precession will have continued until the brightest possible "north star" ever will be in position....the brilliant white *VEGA*, in the constellation of Lyra!

The designation of a DRAGON, or *"water dragon"* to Draco is due probably to two aspects: 1) its brighter stars meander through the sky in string-like fashion that can easily be imagined as the long neck and tail of the dreaded creature; and 2) more importantly is the ancient **legend of the Nile** of which a "dragon" is very important to the people of Egypt.

The Nile river - and hence all rivers pretty much - was life-giving to the ancient Egyptians, its springtime floods providing the irrigation necessary to sustain crops for the growing population. The clouds of the brilliant late spring and summer Milky Way were envisioned by these people as a great celestial river, along which the SUN GOD *RA* would ride in a boat on a daily course through this great river. His head was the mighty sun which shone in full brilliance and majesty when directly overhead. In the Egyptian hieroglyph below you can clearly see *RA* in the center of the river boat, wearing the brilliant sun atop his (its) head. The identities of the other curious features (except the stars surrounding the boat) are unknown.

RA The Sun God
on his journey down the Milky Way

Occasionally, the great *RA* would be overtaken and devoured in this journey by a huge Water Dragon and the bright sun would disappear from sight (this was the solar eclipse in case you are wondering). But much to the credit and strength of mighty *RA*, the sun would ALWAYS pop forth and emerge from the belly of the great Water Dragon.

At any rate, the importance of this star - THUBAN - in the "water dragon" is even more compounded when we consider that it appears that the Great Pyramid of Khugu at Gizeh - the most perfect giant structure in the world - was constructed in such a way that a shaft in the pyramid is aligned **precisely to Thuban** on or about 2,830 B.C. The precision of this alignment of this incredibly narrow and extremely long shaft with what was then the Pole Star appears to be anything but coincidental.

In our GO TO Guide, we also look at the few objects in **Ursa Minor**, the "lesser bear" to its huge and perhaps more interesting "greater bear", Ursa Major.....and to the realm of King **Cepheus**, home of the original and world-famous first Cepheid variable star!

chart shown below can be used as a complete star reference chart for these three constellations. Each of the reference numbers apply to the concise listing of objects found following. The outlines that portray Draco can easily be imagined into the shape of our slithering water dragon....on the other hand, much imagination is required to see a "lesser bear" in Ursa Minor and more so a "king" in Cepheus.

Just for the record, Cepheus discovered - or suspected in some mythology - his beautiful and young wife *CASSIOPEIA* as having some illicit dealings with the handsome, strong and heroic *PERSEUS*. In a fit of jealous rage and resentment, King Cepheus banished the Queen eternally to the sky, strapped in her heavenly throne and perpetually encircling the pole, as the desperate Perseus follows in pursuit in the exact opposite part of the sky. Looking at a star chart or planisphere reveals quickly that, no matter how hard the poor chap tries....he will NEVER catch up with the shackled Cassiopeia.

And....watching the merry-go-round of love from afar is ol' *King Cepheus.*

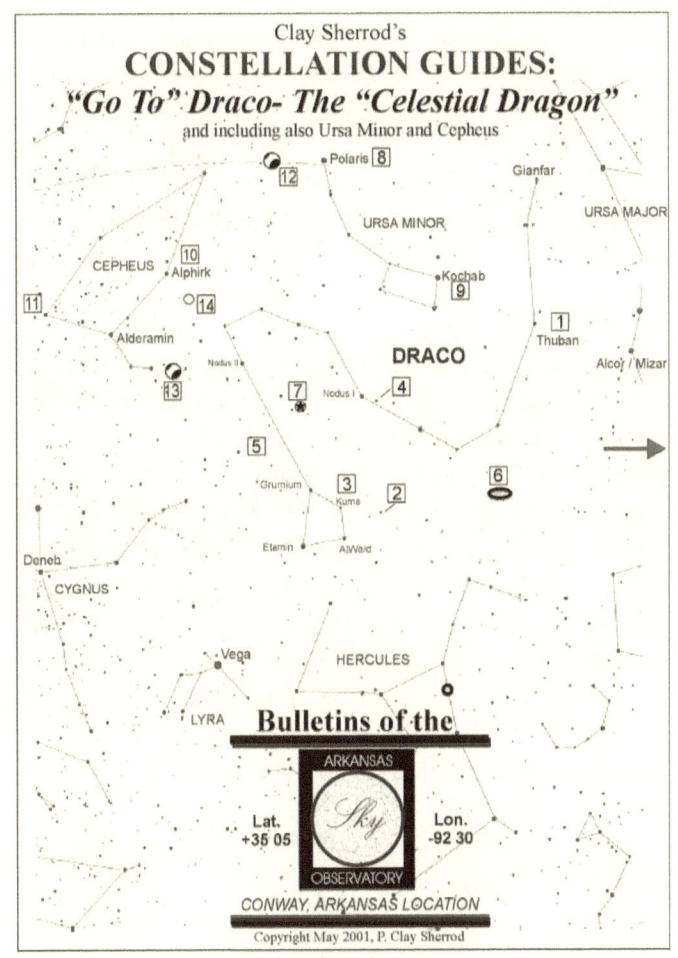

Clay Sherrod's
CONSTELLATION GUIDES:
"Go To" Draco- The "Celestial Dragon"
and including also Ursa Minor and Cepheus

Bulletins of the

ARKANSAS
Sky
OBSERVATORY

Lat.
+35 05

Lon.
-92 30

CONWAY, ARKANSAS LOCATION

Copyright May 2001, P. Clay Sherrod

YOUR DRACO / URSA MINOR / CEPHEUS CONCISE DIRECTORY

There are 14 objects in this constellation GO TO tour; all are in reach of every telescope. Yet each telescope will demonstrate uniquely different and challenging aspects of the objects. In addition to the 14 finest objects, there are literally hundreds of wonderful double stars that are visible in most telescopes as well as many fainter globular clusters (check your AutoStar library for "ngc" listings!), stars of curious colors and motion.

As will all of our GO TO tour guides, I continue to recommend good a good star atlas and/or chart/or computer sky program which lists the finest objects constellation-by-constellation; if you cannot access any of these objects (or those that are not listed in this TOUR), you can access directly from coordinates - Right Ascension (RA) and Declination (DEC) of any known object via the AutoStar or computer tool bar. You merely need to hold down your MODE key on the AutoStar for three (3) seconds and the RA and DEC coordinates appear for the telescope. Merely press "GO TO" and the cursor appears prompting you to enter the Right Ascension of the object if it is NOT listed among the objects in the leypad library; once the RA is entered, press "Enter" and the cursor once again prompts for the Declination coordinates (these coordinates for epoch 2000) are found in all good observing guides). REMBEMBER that all PC, sky program, and Apps use different keystrokes....read the HELP menu!

The following constellation guide to objects at the end of this TOUR will describe all the details of each object and provide specifics as to visibility of that object in YOUR telescope aperture.

The constellation tour star chart provided above (click on and print to size, above) will get you started, as it demonstrates the relative positions of all objects in this "tour" to the conspicuous stars outlining the distinct figures of our three constellations of this TOUR.

Following is a concise directory of the complete 14-object list for your "GO TO TOUR" of Draco, Ursa Minor and Cepheus; you may wish to find the majority of the objects from the AutoStar or computer library (for example, you can merely pull up Messier 102 by going to "Object/Deep Sky/Messier Object/M-102....enter....GO TO" or...if you want to experiment and be a "better computer user", try entering the following coordinates (provided in the list directly following) as described under MODE above. (programs will vary!)

OBJECTS OF INTEREST IN DRACO

OBJECT 1: bright star - THUBAN(alpha Draconis) - R.A. 14h 03' / DEC + 64 37 - Magnitude: 3.6

OBJECT 2: great double star - Mu Draconis - R.A. 17h 04' / DEC + 54 32- Magnitudes: 5.5 & 5.5

OBJECT 3: tough triple star - KUMA - R.A. 17H 31'/ DEC + 55 13 - magnitudes: 4.9(2) AND 7.1 (2)

OBJECT 4: Test for 8-inch - 20 Draconis - R.A. 16h 56' / DEC + 65 07 - Magnitudes: 7.1 & 7.3 (tough

OBJECT 5: one of the closest doubles - ADS11632 - R.A. 18h 43'/ + 59 30 - Magnitudes: 8.9 & 9.7

OBJECT 6: galaxy - Messier 102 (ngc5866) - R.A. 15h 05' / DEC + 55 57 - Magnitude: 10.8

OBJECT 7: planetary nebula - (ngc6543) - R.A. 17h 59' / DEC + 66 38- Magnitude: 8.5

OBJECTS OF INTEREST IN URSA MINOR

OBJECT 8: north star - POLARIS - R.A. 01h 59'/ DEC + 89 02 - Magnitude: 2.1(v)
OBJECT 9: reference star - KOCHAB - R.A. 14h 51' / DEC + 74 22 - Magnitude: 2.0

OBJECTS OF INTEREST IN CEPHEUS

OBJECT 10: nice double - Beta Cep (ALPHIRK) - R.A. 21h 28' / DEC + 70 20 - Magnitudes: 3.3 & 8.0
OBJECTS 11: famous variable! Delta Cep - R.A. 22h 27' / DEC + 58 10 - Magnitude: 3.9 to 5.1 (5.37 days)
OBJECT 12: star cluster - (ngc188) - R.A. 00h 39' / DEC + 85 03 - Mag.: 9.3
OBJECT 13: large open cluster - (ngc6939) - R.A. 20h 30' / DEC + 60 28 - nice, 80 stars
OBJECT 14: diffuse nebula - (ngc7023) - R.A. 21h 01' / DEC + 67 58 - Magnitude: 7.2

A VISUAL GUIDE TO OUR OBJECTS IN THIS CONSTELLATION GROUPING –

Object 1 - Bright Star THUBAN (alpha Draconis)
From the Arabic, this magnitude 3.7 star is actually a translation of the name (Dragon) of the entire Draco constellation. The easiest way to locate Thuban is look for the brightest star almost exactly one-half way from the bowl of the "little dipper" (use the star Kochab) to the famous double stars Alcor and Mizar at the "crook" of the handle of the "big dipper." other than its history of 5,000 years past, the star is most uninteresting other than the fact that it MAT be slightly variable, fluctuating between 3.7 and 4.2 (maybe) and is also known to be a spectroscopic binary (a double star that cannot

be visually resolved, but shows to distinct spectral "fingerprints" with a spectroscope).

Object 2 - Very Easy Double Star - Mu Draconis (ERAKIS)
This double star - at a separation of 3.2" arc - is an easy target for all of our scopes. In the smallest telescopes high magnifications (about 50x per inch minimum) are required to split this pair. Look for two EQUAL magnitude stars (5.1 each) aligned in a nearly precise north-south orientation. It makes a nice average brightness double for the 4- to 6-inch at about 100x. About 14" arc distant is a third small and faint (13th magnitude) companion that MAY be glimpsed with the 8" scope on a very steady and very dark night at about 220x magnification. Averted vision is required to see this companion." This companion star is a very small dwarf star, less than 180 times dimmer than our own sun.

Object 3 - KUMA (ADS 10728) very wide an interesting quadruple system
Okay, now that I've gotten you attention, the main TWO stars are very wide apart - some 61" arc, larger than Jupiter appears at the same power you observe these twin magnitude 4.9 stars! Thus, they are extremely easy in even the smallest scope at low power. However, that being said, there are two MORE companions to this interesting group, now nearly due north of the bright pair. At magnitudes 7.2 and 8.2, and only about 1.2" arc from one-another, the pair can first be seen as one object close to the brighter stars; THEN increase the magnification on the 6- to 8-inch scopes until resolution is achieved. Even though both scopes

can clearly resolve at this level, it is not so easy as it might seem.

Object 4 - 20 Draconis - A very tough double star resolution test for 6- and 8-inch telescope!
This one is tough, and like the object below (Object 5), you will be required to enter the coordinates manually to access it. At magnitudes 7.1 and 7.3, be sure that you have the correct image (it will look like one star at low power) centered in the scope prior to trying to resolve. There are a LOT of 7th magnitude stars in this same area! Once there, begin increasing power on the 6-inch; at about 220x, you should begin to see some clear dark separation in these two stars. Save your time with the smaller scopes....it can't be done.

Object 5 - ADS 11632 - One of the closest double stars to Earth
Being so close to the Earth - only 11.3 light years - this star pair has moved an incredible 30" arc (2/3 the size of Jupiter) in the last 30 years! This is a very interesting pair of red dwarf stars of magnitude 8.7 and 9.7 separated by about 12" arc. This distance is adequate for a clean "split" with the 3-inch and larger scopes, but the relative faintness of them makes this a very tough target and a true observer's test in the 3-inch (and sometimes in a 4-inch). You MUST use your ENTERED COORDINATES AND GO TO on this one....hold down "Mode" for three seconds until you see the RA and DEC coordinates of where you are pointing with the sky program and press "Enter." Then type in the RA as the cursor prompts and once done again press "Enter"; then type in the DEC

coordinates. Once complete, hit "Enter" again, and the telescope will take you there!

Object 6 - Messier 102 (ngc5866) - A very nice and unusual galaxy

Although known as "Messier 102" this object was likely NOT seen by Messier and not originally part of his list; several additions were made after his list was published of objects that more modern astronomers "thought" that Mr. Messier should have seen (or perhaps "did" see) and thus they felt it necessary to add them to his famous list of 100. Messier 102 is an object worth attempting to locate. At magnitude 10.8 or so, it is one of the most difficult galaxies on the Messier list, but it is so small that its brightness is "compacted" into a more brilliant center. In photographs, this object looks like a "flying saucer" seen from the edge, as can be seen in the beautiful Lick Observatory photo below. Because of its concentrated brightness, it is possible to see this galaxy and a bit of this shape even in the 4-inch telescope at about 120x....the "UFO shape" is clearly evident in the 6-inch and larger telescopes at about the same magnification. This faint object is just on the verge of visibility to the smaller telescopes at best will appear as a very faint out-of-focus star.

IN URSA MINOR.....

Object 8 - Our "North Star" **POLARIS** (alpha Ursa Minoris)
What can I possible add to the already voluminous accounts of this otherwise uninteresting star? Right now, it just happens to be our north polar star (see above), or otherwise it would be a very often-overlooked medium bright star will little to offer casual observers. But that has not stopped civilizations throughout history from building great temples, churches, monoliths to its honor. Indeed, in earlier times, Polaris was known as "*Al Kutb Al Shamaliy*", or the *northern spindle* and signified the beacon toward mighty **Mecca**. For observers today, Polaris IS a wide **double star** and one worth taking a look at! It is listed on your sky program, so why not? Have you ever done a GO TO to the object you usually use for home position alignment? I doubt it. The brighter member of the Polaris pair is a variable *Cepheid* (see discussion below) of magnitude 2.3 normally. The faint 9th magnitude companion of Polaris is a whopping 19" arc distant, about due west of Polaris itself. It is really worth a look, as this tiny star in the 6-inch is a distinct BLUE COLOR and quite a nice contrast to the

brighter star. Use medium (about 75x) magnification for the best look for color and contrast.

Object 9 - Our famous KOCHAB (beta Ursa Minoris)
About 3,000 years ago, Kochab was very near the north celestial pole, so it has shared this honor with Thuban and Polaris. This star is referenced here because it has become somewhat "famous" with telescope users from the increasingly used "Clay's Kochabs's Clock method of accurate portable telescope polar alignment first published on ASO; the Kochab Clock is available from the ASO Home Page tabs at top; use it: you will love it!

IN CEPHEUS

Object 10 - Nice Double Star - a great test for small telescopes - ALPHIRK (beta Cephei)
Here is an excellent challenging double star for small APO refracting telescopes. *Alphirk* (also spelled "ALFIRK") is relatively bright (magnitude 3.1) star with a fainter 8th magnitude companion some 14" distant, nearly due west from the brighter star. It is an interesting object that will require about 100x in these scopes to see the fainter star. In the 4-inch it is a very nice low power target, but loses some of its beauty in larger instruments. It might bear looking at regardless, however when you know an interesting aspect of this very bright "newer" star. At a distance of 930 light years, it has an actual luminosity that - if it were located by Earth at the same distance as our sun....it would OUTSHINE it by 4,000 times!

Object 11 - Delta Cephei - the "granddaddy" of the **"Cepheid Variables"**
By nature of its light changes of ONLY 3.6 to 4.3 magnitude and its AVERAGE brightness of only 3.6 or so, this star should be relatively uninteresting. However, you can literally set you clock by the variations in light of Delta Cephei, with its precise pulsating period of **5.366341** days!

If you are interested in observing this star (naked eye only....there is nothing to be gained by using a telescope to estimate brightness changes) you can download and save the file the following link for a chart of Delta Cephei; merely save it to file, resize and print off the chart which shows a wide field of view and many comparison stars suitable for comparing through Delta's entire cycle. Link to this chart at:
https://www.aavso.org/apps/vsp/ . Note for these charts, simply type in the NAME of the variable at top to generate your choice of chart.

This huge yellow giant "K" type star was first discovered as variable by **John Goodricke** in 1784 and is perhaps the "most watched" variable outside of *Algol*. These stars are now quite common and have been found to actually GROW and SHRINK in size, accounting for their apparent magnitude variations. With periods of merely a day to some 50 days, the Cepheids can be used for incredibly accurate gauges for determining distances to objects in space, merely by their apparent brightness and their periods of variation. I highly recommend to all observers to take the time to reference these interesting stars and read of their history and usefulness in modern astronomy.

Object 12 - "Ancient" Galactic Cluster - ngc188
Located only 4 degrees SSE of Polaris, this very distant (5,000 light years) galactic cluster may well be on the order of 24 BILLION years old, one of the oldest known. The nature of its stars suggests that it is more similar to the GLOBULAR CLUSTERS, like Messier 13, than to the loose and younger "galactic star clusters". IMPORTANT NOTE: if this estimation of the age of this cluster is, indeed accurate....it makes it perhaps the OLDEST known object, not just in our own galaxy....but in the entire UNIVERSE.

With 6- and 8-inch telescopes, this cluster is impressive, appearing as a "dusting" of tiny stars just at the threshold of visibility. Only a few of the brightest of its yellow giant stars can be detected. At magnitude 9.1 and covering a large 15' (one-half moon diameter) of sky, this object is too faint for small telescopes and is a quite difficult object even for the 4-inch. With the 6- and 8-inch scopes at very low magnification, it is an awe-inspiring sight, with the faint stars around the edge of M-62 glowing clearly and the bright cluster suspended in the starry field.

Object 13 - Guess what...ANOTHER galactic cluster: ngc6939
This is yet another star cluster, but this one is much richer in stars (at least that can be seen) and not so distant as ngc188 above. There are about 80 stars in this cluster, most of which are visible in this VERY tight (only 5' arc) cluster with the 8-inch at medium-

high power. At about 125x in the 4- and 6-inch scopes this cluster can reveal about half that many. this cluster is beyond the reach of smaller telescopes.

Object 14 - Diffuse Reflection Nebula - ngc7023
This is a rather large but overall very bright (magnitude 7.3) nebula due southwest of Alphirk. Because it is spread out so much (over 18' arc of sky - "2" Jupiter's), it requires a VERY dark sky to see this in all scopes. Use as low a power as possible and do not attempt to observe unless sky conditions are almost perfect. Low powers and very wide field reveal a very interesting and memorable field of stars..

WANDERING ABOUT....YOUR NEW "USER OBJECT" IN CEPHEUS

This brief GO TO Tour of Draco, Cepheus and Ursa Minor has taken you into one of the many "void" areas that seem to not have a rich display of astronomical delights. But remember....all the cosmos should delight you! THIS USER OBJECT is not even part of our TOUR this time! It's like a Christmas surprise, and the COLOR is quite appropriate for such an occasion as well. We are going to load as your next Telescope User Object:

"Herschel's Garnet Star" - R.A. 21h 42' ; DEC + 58 33. Magnitude 3.7-5.0 (variable). Here are the details on this remarkable star. It is a variable with MULTIPLE superimposed periods, and thus the true period is NOT known. It is an incredibly "late" highly evolved giant star and very, very red! Hence "*Garnet Star*". You will be thrilled with this star at low to medium power, appearing as a ruby in the sky! This is a gem for all of our telescopes, so load this one up!

LET'S LOAD THE "GARNET STAR" AS A USER OBJECTS! You could not find any more appropriate and interesting subject to add to your growing "User Object" library.

On your PC (use instructions for tabs and keys) or AutoStar, go to: "Select/Object [enter] and scroll down to "user object" [enter]. Now enter the coordinates listed above for the "Garnet Star". Under "description" put in a nice title for it ...be creative with your brief description.

*   *   *

"Even from the abyss of horror in which we try to feel our way today, half-blind, our hearts distraught and shattered, I look up again and again to the ancient constellations that shone on my childhood, comforting myself with the inherited confidence that, some day, this relapse will appear only an interval in the eternal rhythm of progress onward and upward....."

Stefan Zweig

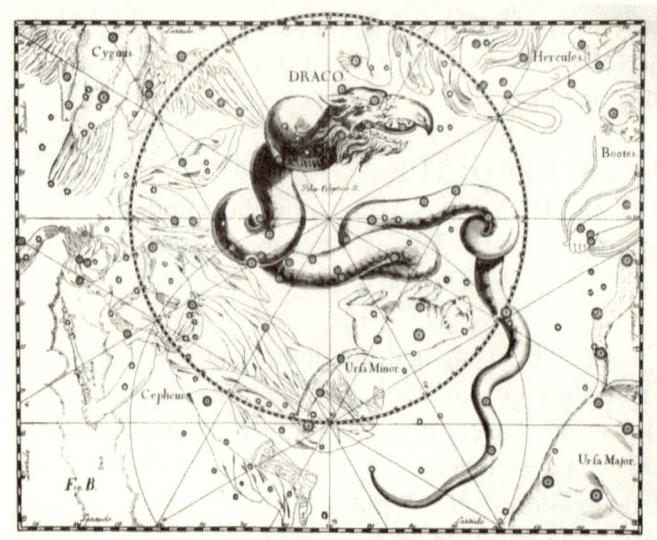

Hevelius' Draco 1687

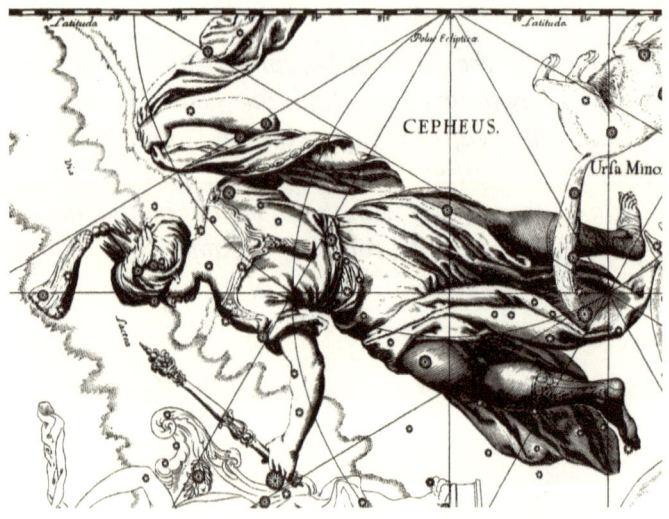

Cepheus by Johannes Hevelius, 1687
Note the convergence to the North Celestial Pole
At top

CHAPTER 15

ERIDANUS

....among the sky's bright and beautiful stars: "A River Runs Through It"

As our *Constellation* guides take us ever so deeper into the rich and clear winter skies of the northern hemisphere, we embark on a float trip down the Celestial River ERIDANUS in our 31st installment. It is appropriate as we approach the half-way point in these TOURS that we have discovered the "source" of so much watery references throughout our fiction-filled firmament. In this watery abyss that is space, we find a sea monster or whale (*Cetus*), and a Dolphin (*Delphinus*), not to mention both a large and small water snake (*Hydra* and *Hydrus*), the leaping swordfish (*Volans*), a guy carrying water around (*Aquarius*), a ship's sail (*Vela)*, apparently a fishing net that got lost from the same boat (*Reticulum*), along with their compass...no wonder they lost their ship! (*Pyxis*), a northern fish (*Pisces*) and southern fish (*Pisces Austrinus*), a water crane (*Grus)*, a swan in flight over the water (*Cygnus*), yet ANOTHER compass (*Circinus*), the water crab (*Cancer*), and - oh my goodness, we've found the keel from the Argonaut's ship! (*Carina*) and of all things.....a SEA GOAT! (*Capricorn*).

This is not to mention the scores of star names from both north and south celestial hemispheres that were aptly named for or about the oceans, sea myths and mythology, water and water-related creatures. So WHERE does all this stuff reside? Where is the remains of the tattered and torn vessel of the famed

367

Argonauts....the home of the sea monster, and for gosh sakes WHERE does a Sea Goat live?

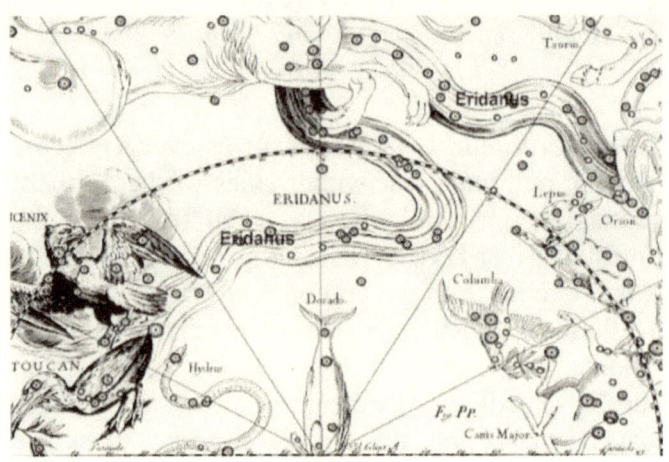

The celestial river, Eridanus
As drawn in 1687 by Johannes Hevelius

Why... in *Eridanus*, of course! Curiously since this constellation is a vague and sprawling with few spectacular and eye-catching asterisms, this was actually one of the 48 original constellations assigned by **Ptolemy** in the second century A.D. From Egyptian influence perhaps more than 2,000 years before him, Ptolemy cast honor up this meandering stream of stars to represent the great river in which the unfortunate Phaethon plunged to his steamy death. After learning that his father was a great god (*Helios*, the sun god), *Phaethon* (a mere mortal) begged his father ("Daddy.... Puuullllease!!??") to allow him one morning the great honor of driving the fiery chariot of the sun across the sky from east to west to bring great heat and light to the mortals below on Earth ("Please, Daddy....PLEASE?!"). Begrudgingly Helios finally

368

gave in like all fathers and allowed the ill-fated day to begin with Phaethon and his flaming cargo rising as predicted in the east and begin its path across the sky....but unfortunately VERY recklessly (something here about a "teenage driver," but it was lost in the translations). Soon, Phaethon lost control of the mighty chariot and sure enough it began careening out of control, surely to whack the Earth in a fiery fireball of doom.

Since *Zeus* (as we have seen) had many love interests in the beautiful mortal women of Earth, he could not allow such a catastrophe to lay waste his harem and thus, cast a huge thunderbolt upon the descending Phaethon and his chariot, sending in plummeting instead into the perpetual River Eridanus, which steam perpetually (the Milky Way) from the now-extinguished sun. The story does not say anything about WHERE Helios was able to find another sun in time for sunrise the NEXT morning, but I have it on good authority that Wal Mart was having a "Sun-day" sale.

The steaming mists from this still simmering sun fill the winter skies even today, as the rich winter Milky Way to the east of River Eridanus carries the clouds far to the north, through Orion, Taurus, Auriga, Perseus and Cassiopeia.

It is interesting to me that even the constellation CYGNUS, the swan of summer skies (also in the Milky Way) originated in this story as well.....it so happens that a very devout friend of Phatheon, seeing his plight as he plunged into Eridanus, dove in after him in an ill-fated attempt to rescue the young man. He drowned and showing compassion for so great and trusted friendship, Zeus changed the lifeless body of his would-be rescuer into the form a beautiful swan to fly above the river of the sky for all eternity.

This tale did NOT originate in ancient Greece, however, even though that is how we remember it best. The legend of the sun-toting chariot goes far back to the ancient Egyptians where the Sun god, Ra, would rise steadfastly in the east and cross the great river (the Milky Way) of the sky each day and disappear at the end bringing darkness once again to the world.

Cool story, huh? I hope you enjoyed that, because the rest of the constellation is large and empty. Just kidding of course, as Eridanus is filled with wonderful double and multiple stars, and at least two dozen galaxies that are in reach of many amateur telescopes (most 8" and larger). There is not ONE galactic cluster in Eridanus that is viewable in instruments smaller than 24 inches. Only one planetary nebula (ngc1535 discussed below) is within reach in this huge constellation, and there are but two very faint and diffuse nebulae, both of which are so large and spread out that they are but invisible except in wide field extended photographs.

Yet this constellation is by far the LONGEST south-to-north in all the sky. Just look at the chart below: you note that Eridanus' northern boundary rest exactly on the celestial equator...."0" degrees declination. You can trace the boundary of this celestial meandering river all the way past the 9th brightest star of the sky (*ACHERNAR*) to nearly

MINUS 60 degrees south latitude! Sixty degrees of VERY sparsely populated sky connecting observers in New Jersey and South Carolina in the northern skies to those in New South Wales in the southern!

Clay Sherrod's
CONSTELLATION GUIDES:
"Go To" ERIDANUS : the "Celestial River"

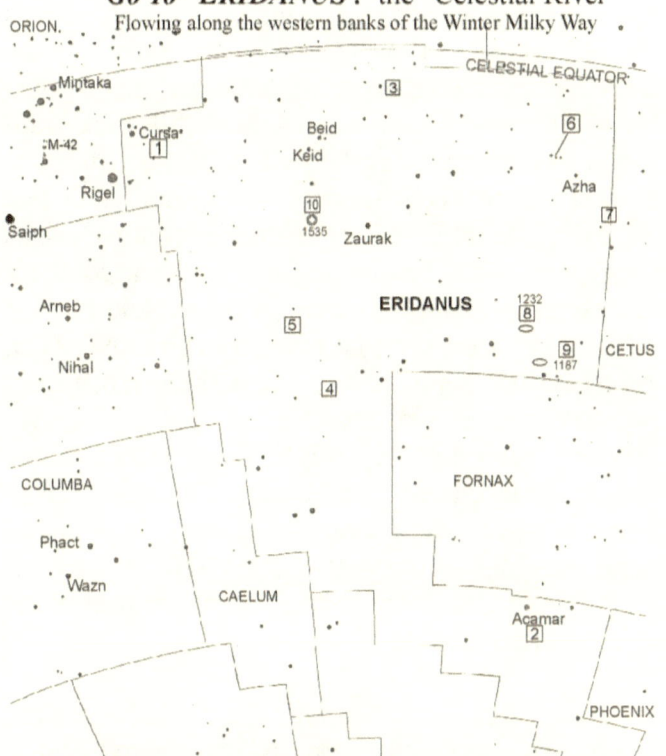

A finder chart for locating many of the GO TO objects in the constellation of Eridanus; if using a computer planetarium program, you are encouraged to plot the objects on your screen for higher resolution than this chart provides.

*   *   *

371

THE STARS OF ERIDANUS –

For a constellation that extends so far south and with little flare to draw attention to itself, Eridanus has a considerable number of "named" stars, and 10 stars of magnitude 4.0 or brighter! To me, one of the most curious aspects of this constellation from the earliest days in regard to the names of its stars is the fact that all stars in Eridanus were named by the same Arabian stargazers of antiquity that named most all other stars.....names that still stick until today. When you consider that *ACAMAR* (theta) is at a far south declination of -40 degrees and that *ACHERNAR* (alpha) is even farther south - minus 58 degrees - it is interesting that such far southern stars were accessible to them in the Middle Eastern world. However, from latitude 35 degrees NORTH, ACHERNAR is just barely below the horizon from my observatory at its highest point.....I have routinely seen this very bright star RIGHT ON the watery horizon of the Gulf of Mexico from Pensacola, Florida on many occasions! On the other hand the star somewhat fainter star Acamar is visible high above the southern horizon from most mid-northern latitude locations.

In the ancient Arab world, with latitudes farther south than most of Europe and the United States, note that the ENTIRE expanse of Eridanus is visible, all the way past its brightest star, Achernar.

NOTE that our most northerly object (in the abbreviated BONUS listing of NGC galaxies found following) is ngc1637, which is located "high" in the constellation at declination MINUS 02 degrees! All objects in this "GO TO" tour thus will have negative declinations (south of the celestial equator) and you are cautioned to note that many references to NOT carry through the "-" (minus sign) ahead of the appropriate declination coordinate.

For a complete cross reference of named stars, Bayer and Flamsteed designations, SAO numbers and double stars along with their coordinates, I refer you to the wonderful
http://www.deepskywatch.com/deepsky-guide.html guide to deep sky objects in all of the constellations.

From mid-northern latitudes, the highest stars of Eridanus - those clustered near declination "0 degrees" - rise at about 9 p.m. throughout April; however, not until mid-January do the same stars appear highest in the sky at the same time. Midnight culmination of the same stars always occurs on about December 3 each year, reaching that peak nearly the same time as the very bright white star in Orion RIGEL, just to the constellation's east.

Eridanus is virtually FILLED with splendid double and multiple stars, well over a hundred of which are observable in much detail in telescopes from 3" to 16". Unfortunately I am able to only select a few of the finest double and multiple stars to discuss in our Eridanus "GO TO" Guide; I very much encourage all telescope users to obtain the three-set copy of *Burnham's Celestial Handbook* for reference on each and every constellation; there is no finer reference work for deep sky viewing to be found.

Also note that good PC programs are now available for you to install on your computer to CONVERT the epoch 1950-1960 coordinates listed in the

Burnham reference tables DIRECTLY to epoch 2000:.....all you must do is merely type in the coordinates exactly as they appear in Burnham's and enter, with the resulting 2001 conversion done instantly for use with your PC or controller.

Once scanning the pages of Volume 2 of *Burnham's*, you will quickly see the "best multiple" and unusual stars for observations. There are no less than 100 fine multiple stars within reach of most amateur telescopes, many of which are fascinating objects. A few have been selected here as described following.

OBSERVING THIS CONSTELLATION WITH BINOCULARS –

For those who wish to explore the regions of Eridanus in binoculars I highly recommend a standard good quality 7 x 50 or 10 x 50 glass used in very dark, moonless skies away from artificial lighting. Remember to let your eyes become "dark adapted" for at least 15 minutes prior to searching out fainter objects. For a wonderful selection of binocular tour objects, visit http://www.dibonsmith.com/eri.htm, a tremendous guide by Richard Smith from his web page entitled "The Constellations."

GETTING STARTED –

As with every "GO TO" TOUR guide, each GO TO object in ERIDANUS is discussed for your telescope regarding the type of conditions necessary for you to view it optimally for discern the very faintest details.........magnifications and aperture necessary for most objects, and much, much more. This is YOUR complete GUIDE to get you on your way to exploring the best (and few!) objects in this HUGE constellation. The chart provided above from the Arkansas Sky Observatory and the subsequent detailed listing of "BEST" objects contains the finest or most interesting from my own observing experience and preference.

Use the attached star chart shown above and the following Guide as an excellent reference for your next star party itinerary, or a beginning for further study into the thousands of objects visible in this part of the sky. To access and print the chart, double click on it and save the image to a file on your computer. Once saved, open the file and RESIZE this image to fit the normal paper format for your program and save again....then merely print out the chart on high quality paper for a field reference in this GO TO TOUR!

OBSERVING TIPS –

Every deep sky object and every double/multiple star will have a "PERFECT MAGNIFICATION".
his is the magnification that you should use that will show the object as bright and with as much as detail with possible and still increase its size appreciably so that you can view it comfortably and unmistakably. The rule for determining "optimum magnification" is that: 1) too low power results in sky background glow detracting or diminishing the contrast against the deep sky object; 2) too high magnification darkens BOTH the sky background AND the object; 3) medium magnification can be achieved at which you have MAXIMUM contrast

between the object and its darkened background sky. I have found through three decades of direct observing that about 15x per inch aperture for deep sky observing is PERFECT for most objects. That being said, always remember that DOUBLE or multiple stars require whatever power you can crank out....the seeing conditions are the limiting factor here.

For my complete and comprehensive discussion regarding seeing conditions and sky transparency, see the discussion under *ASO Guides* on the Arkansas Sky Observatories website.

With all deep sky objects, avoid attempting to observe when the moon is in the sky, even a very thin crescent, as its brightness in the sky will overshadow the very dim contrast afforded by even the brightest deep sky object; if you see the object at all against moonlight, you will NOT see the subtle outlying areas or the full detail of what is presented.

For detail descriptive lists of the great double stars within ERIDANUS, and as with all of the "GO TO" tour constellation lists, I recommend a good star atlas and/or chart which will list all the finest objects, constellation-by-constellation. One very handy reference guide is the *PETERSON FIELD GUIDE TO THE STARS AND PLANETS*, which features complete lists with declinations, right ascensions, magnitudes, and all pertinent information for you to expand your observing horizons beyond this brief guide.. For the many double and multiple stars, I again urge you to refer to the indispensable *"Burnham's Celestial Handbook"*, Volume 2 for a complete abbreviated listing.

Truly these extensive *Constellation* study guides will most definitely put your AutoStar or PC sky program to work for you in the most efficient and enjoyable way possible! As a matter of fact, MANY sky program users are now programming their own "Tours" based on these guides, using each constellation as a separate GO TO tour for the sky library that can be added in or deleted through the main edit screen on your PC or MAC computer.

We hope you enjoy these comprehensive guides to touring the constellations via your AutoStar or its computer-driven telescope. Each new installment is complete with diagrams, charts and illustrations that you will find nowhere else. Please let us hear YOUR feedback and your observations of each and every constellation after YOU have toured its vast reaches of our skies!

YOUR ERIDANUS CONCISE DIRECTORY OF INTERESTING OBJECTS –

As mentioned, Eridanus offers a wealth of fine double and multiple stars (for a full discussion on double star observing and their "Position Angles" refer to my brief overview in the "GO TO" TOUR guide for Lacerta in Vol II of our *Constellation* guides in this publication.

The most interesting 10 targets in the constellation have been chosen for this ERIDANUS "GO TO" tour; as with all guides, all objects listed below will be visible in most telescopes (some naked eye) from 3inches to 8 inches; of course larger apertures may "show" an object a bit closer and "better," but frequently a wide field and low power view is more desirable than aperture for FINDING the objects initially. Indeed, I strongly encourage you first FIND the target object, or its approximate location through your GO TO function with your lowest power and then - once IDENTIFIED positively - move up slowly in steps with magnification if necessary. Remember, not all objects "like" magnification. Sometimes better "field of view" of a small telescope is desired over light gathering and magnification of larger telescopes.

The constellation tour Star Chart above (click on and save to a file on your PC; then open it and re-size to fit the page and print for a very handy at-the-scope star chart) will get you started on your journey for this constellation.

Following is the concise object list for your "GO TO" tour of ERIDANUS; you may wish to find many of the objects from the AutoStar Library (for example) you can easily go to the planetary nebula ngc1535, if you pull up "Object/Deep Sky/NGC Object/..then type in '1535'...." and then press "Enter", followed by "GO TO" to access this nice but small and bluish shell of gas.

You will access your FIRST GOTO target - (usually the brightest star in each constellation) - via the command "SETUP / OBJECT / STAR / NAMED....and scroll to EITHER "ACHERNAR" or "CURSA," then press "Enter" and subsequently "GO TO" to move to one of these bright stars. Having TWO stars as our first GO TO object is unique for Eridanus, because of its huge span north-to-south. Since ACHERNAR is so far south as to render it invisible from mid-northern latitudes, the "Beta" star CURSA has been provided for observers "up north;" for those "down under", Achernar will be a much more inviting first target!

You may also access the constellation by: SETUP/OBJECT/CONSTELLATION/"Eridanus"... ..Enter....GO TO, which will slew your telescope very near the cartographic center of this sprawling star group. (*REMEMBER: different keystrokes for different programs!*)

OBJECT 1: bright stars –
ACHERNAR (alpha Eri) - R.A. 01h 38' / DEC - 57 15 - Mag: 0.53 - Very far south, 9th brightest star!
CURSA (beta Eri) - R.A. 05h 07' / DEC - 05 05 - Mag: 2.8 - A bit more comfortable position for northern hemisphere! ****PLUS double 66 Eridani!****
OBJECT 2: double star - ACAMAR (theta Eri) - R.A. 02h 58' / DEC - 40 18 - Mags. = 3.5 & 4.5 - very bright and wide white pair, excellent for all scopes!
OBJECT 3: nice double - 32 Edi - R.A. 03h 52' / DEC - 02 57 - Mags. = 5.1 & 6 - very nice star, good color contrast. Great for all sizes.
OBJECT 4: quadruple star - B744 - R.A. 04h 22' / DEC - 25 44 - Mags.= 7.3, 7.5, .5, & 12 - Neat star for larger scopes!
OBJECT 5: test double for 3" - B184 - R.A. 04h 28' / DEC - 21 30 - Mags. = 6 & 7 - very close, 1.6" arc. Good ETX 90 test star.
OBJECT 6: test double for 5" - 9 Eridani - R.A. 03h 03' / DEC - 07 41 - Mags. = 5.5 & 10 -

Excellent test for 5", with 1.9" separation and very faint companion.

OBJECT 7: variable star - Z Eri - R.A. 02h 48' / DEC - 12 28 - Mag. range 6.5 to 7.9 - Very nice variable for small scopes/binoculars!

OBJECT 8: spiral galaxy - ngc1232 - R.A. 03h 10' / DEC - 20 35 - Mag. 10.4 - very nice with much detail in larger telescopes! Face-on galaxy.

OBJECT 9: spiral galaxy - ngc1187 - R.A. 03h 03' / DEC - 22 52 - Mag. 11.2 - large, fairly faint, but two spiral arms visible in larger scopes.

OBJECT 10: planetary nebula - ngc1535 - R.A. 04h 14' / DEC - 12 44 - Mag. 9, nice small bluish object with 11th mag. central star.

*** BONUS ***

OBSERVABLE GALAXIES IN ERIDANUS - Faint galaxies that are within reach of many common amateur telescopes:

NGC (min. arc)	R.A.	DEC.(minus)	MAG.	SIZE	TYPE	DESCRIPTION
1084	02 46 (-)	07 34	11.0	2.1 x 1.1	Sc	Fairly bright, possible in 5", elongated and small
1187	03 03	22 52	11.2	5.7 x 3.7	SB	Barred spiral, very large but dim, oval shaped; easy in 8" difficult in smaller scopes
1232	03 10	20 30	10.4	7.0 x 5.5	Sc	Very large and fairly bright, round. Can be seen in 3" scopes, but very vague
1291	03 17	41 06	10.1	5.0 x 2.0	El	Very nice elliptical, larger than most in area; oval shaped and visible in 3"

1300   03 20   19 24   11.1   5.7 x 3.5   SB   Large barred spiral, should be seen in 5"; cigar-shaped smudge

1313   03 18   66 29   10.7   4.5 x 3.0   Pec   Very nice in 5" and larger, nearly round and larger than most

1309   03 22   15 24   11.4   1.9 x 1.7   Sc   Elongated spiral, fairly small and indistinct; requires 8" or larger for sure

1325   03 25   21 32   12.2   4.2 x 1.1   Sb   Nice but very faint, 8" and larger required; very thin wisp of light

1332   03 26   21 21   10.4   3.4 x 1.0   El   Nice brighter elliptical, very elongated and brighter toward center; starlike in 3"

1353   03 23   20 50   12.1   2.5 x 0.9   Sb   Very thin and faint spiral, nearly edge-on;   will   require   larger   telescopes

1357   03 33   13 40   12.3   1.4 x 0.9   S   Tiny spiral, looks like a very small oval "ghost" in 8"; 12" will show clearly

IC1953 03 34   21 29   12.3   2.1 x 1.9   Sc   Very faint and small spiral with no detail; glimpsed in 8"

1359   03 34   19 31   12.4   1.6 x 1.3   SB   An even dimmer barred spiral that is very small and difficult in all scopes

1386   03 37   36 00   12.2   2.5 x 1.0   S   Narrow streak of very faint light, just a smudge in larger scopes

1395   03 39   23 01   11.1   2.1 x 1.4   El   Fairly bright and small elliptical; easy target in 5" but nearly starlike

1400   03 40   18 41   10.7   0.8 x 0.7   El   Extremely small elliptical but great target for 3"; starlike in all scopes

1407   03 40   18 34   10.6   1.1 x 1.1   El   Very small but brighter elliptical; starlike

1421   03 43   13 31   11.3   3.0 x 0.6   Sb   Nice large and narrow galaxy, nearly edge-on; difficult in 5", distinct in 8"

1518   04 07   21 10   11.9   2.4 x 0.9   Sc   Very thin and faint spiral, like a transparent pencil in the 8"; visible in 5"

1537   04 14   31 33   11.8   1.2 x 0.6   S   Very small and difficult spiral even in 8" scopes

1600   04 32   05 04   12.1   0.8 x 0.6   El   Very tiny elliptical, starlike in 8"; may be glimpsed as a test in the 5" scopes

1637   04 41   02 50   12.1   2.6 x 1.9   Sc   Medium sized spiral but very faint; 5" will show it,but 8" and larger preferred

1640   04 42   20 26   11.9   2.0 x 1.1   SB   Barred spiral that is distinct in 5" and interesting in 8" and larger scopes

1700   04 57   04 31   11.9   0.9 x 0.8   El   Tiny elliptical requiring at least the 5" telescope

YOUR VISUAL GUIDE TO DEEP SKY OBJECTS IN ERIDANUS

Object 1 - Our "Starting" Bright Stars –
"ACHERNAR" (alpha Eri - pronounced: "ATCH-er-nar") / CURSA (beta Eri - pronounced: "KER sa")
** ALSO - double star "66 Eridani **

The brightest star of Eridanus, 0.6 magnitude ACHERNAR reigns as the 9th brightest stars of the skies encircling the Earth....yet most of the worlds population has never seen it. At latitude -58 degrees as stated above it never reaches high enough above the southern horizon to be seen anywhere north of about north latitude 30 degrees. However, for our observers "down under" this is a benchmark star and a beacon of late fall skies. From the Arabic "Bright one at the River's End", this star name is one of my favorites, signifying the long 60-degree run from north to south of our celestial river, Eridanus. We can only imagine ancient mariners skimming the waters of the southern seas using this bright star as their guide. On about October 20th, the star "might" be seen at midnight immediately over the waters of the Gulf of Mexico from Florida, Louisiana and Texas as well as from all of the Hawaiian islands. It is not visible from the Mediterranean countries. Achernar is an extremely "early" type bright white star that is quite larger than our own sun.....glowing brightly at a distance of only 120 light years.

Nowhere near as bright, but still conspicuous for more northerly observers is Beta Eridani, CURSA, shining at a seemingly pale magnitude 2.9. However, this star is VERY close to our solar system (80 light years) and more evolved than is Archernar. Cursa is best found as the brightest star northwest of the distinct RIGEL in the "feet" of Orion. NOTE: look for the fine wide double star 66 Eridani just to the WEST of Cursa, easily spotted in the same very wide field of view with your telescope; 66 ERI is a true double with a 6th magnitude primary star and a 9th magnitude companion about 1 arc minute due north, a wonderful and easy target for most telescopes.

Object 2 - The Fine Double Star ACAMAR (pronounced "ACK-mar") - (theta Eridani)
Although this star is VERY low in the southern skies, it still can be seen by observers in the southern United States and other locations south of north latitude 35 degrees. This is an interesting star, both from its brilliant white appearance of its magnitude 3.4 and 4.4 stars, but also because both stars are moving uniformly through space without any signs of orbital motion around one-another! Separated at about 8.3" arc, this pair has been so observed since John Herschel began accurately measuring their Position Angle and separation early in the 19th century from South Africa. The Greek **Ptolemy** cataloged this star and ranked it among the brightest, or his "of the first magnitude." Today, the star is almost 3rd magnitude so we are witnessing a star that likely has actually FADED in luminosity in recorded history. Nearly the same spectral type star as Achernar (see above) and at nearly the same distance (only five light years closer than the bright Achernar), it is likely that both these stars are fairly new stars.....with Achernar a blue-giant and Acamar and white, slightly more evolved star. Hence its slight age difference might account for its dimming from simple nuclear stabilization, whereas Achemar (a "B" type star) is still very much in its formative stages, much like the bright blue stars of the Pleiades star cluster.

Object 3 - A Colorful Double Star - 32 Eridani - Nice Object for All Scopes

This is a BEAUTIFUL double star, with a wide separation of 6.9" and thus resolvable in all scopes larger than 3". The primary is an orange-yellow 5th magnitude star, while its companion is a 6th magnitude star of brilliant white color. So very near the celestial equator at only minus 03 degrees, this is a must-see for telescope users. At present the stars are oriented in a nearly north-south direction

Object 4 - A Challenging Quadruple Star! "Burnham 744"
(see my discussion on double star positions and observing in the *GUIDES/Constellations/Lacerta*.
This is the 744th multiple star measured carefully and cataloged by the famed **S.W. Burnham**, the father of American double star observers. His keen eye and carefully measured stars have served as the benchmark for our observations today, noting the orbits, periods and even the masses of thousands of double and multiple stars. B744 is an unusual Quad Star, the primary component being a PAIR of equal brightness magnitude 7.4 stars. At present, these stars are as far apart as they have been since their discovery in the 19th century. Although the Position Angle (see chart below) is not appreciably changing since its discovery, the distance between these two very close stars is increasing. Right now the stars measure some 0.9" arc apart, theoretically far enough apart for resolution with a 5" telescope; however, I have had considerable difficulty resolving this star, partly because of its very southerly position (-26 degrees declination) and thus unsteady air currents between my telescope and the star. An 8" telescope resolves it cleanly into a neat yellow pair.

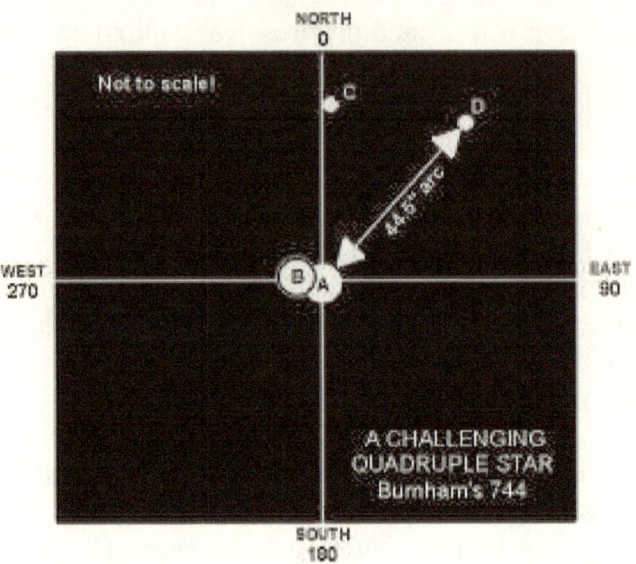

Not to scale!

NORTH
0

C

D

44.5" arc

WEST
270

B A

EAST
90

A CHALLENGING
QUADRUPLE STAR
Burnham's 744

SOUTH
180

Observers south of 35 degrees north latitude are encourage to attempt to resolve this star, and report what telescope, seeing conditions, magnification required if successful. HOWEVER...don't stop there! There are two more parts to this star! The chart above shows the 12th magnitude "C" companion a wide 39" arc due north of the difficult "A" and "B" pair; this will require at least a 5" telescope simply because of the faint magnitude. NOW....look almost the same distance (but a little farther) as that faint star is from the A-B pair almost northeast (Position Angle 41 degrees) to find the "D" component, a much brighter star 8.5 magnitude, a good target for a 3" scope. Actually, if you will find the "D" star first and then look for the "C" star, you will have a better idea of how far that one is from the brightest pair. Thus, in one star, we have a Quad star that works as follows: TWO stars (the A-B combo as one star and "D") can be seen with a 3" scope; THREE stars (A-B combo as one star, "D" and "C") and all FOUR stars with an 8" or larger scope. Keep in mind that the keen-eyed observer with a 5" scope, good steady skies and high magnification CAN split the "A" and "B" components!

Object 5 - An Easier Test Double for the 3" and 5" Scopes - Burnham 184 (B184)

Can your 3" scope hit Dawes' limit? Here is a nice test to find out. B184 is a pair of magnitude 6 and 7 stars, with the fainter star 1.5" arc away from the main star in Position Angle 252 (nearly due west). A 3" scope should be able to resolve two stars of equal magnitude under steady conditions that are 1.3" arc apart. Hence, this one should be tough, but do-able. If the air is steady, use as high magnification as you need; you will first see the point of light "elongate" in an E-W direction with high power and finally separate with some nice dark sky between them with even higher power. NEVER hesitate to use high magnification on nights of very good steadiness. For a complete discussion regarding how "seeing" and air transparency affects your viewing, go to my discussion in the ASO Guides - Frequent section of this website . I routinely use over 50x per inch aperture to observer planets and doubles stars when the conditions are favorable; the old myth that higher powers are useless on telescopes are not true with today's Null-figured optics and excellently corrected eyepieces.....it is the AIR around us that limits the "power potential" of your telescope.

Object 6 - Now...a Test Double for a 5" Telescope! - "9 Eridani" (also Rho Eri)

There are actually THREE stars that comprise the designation "Rho Eridani", with the Flamsteed numbered star "9" being the middle of the three; actually they comprise a very pretty small triangle in the very northwestern part of the constellation, and hence are very accessible for northern viewers. Our double star is the southern "apex" of this triangle. This star is steadily closing and at present there is about 1.7" arc separating "A" from "B" (see chart below). The brighter star (A) is magnitude 5.5 and hence very easy to spot as the middle of the three "Rho stars." However, it is the faintness of the "B" star (magnitude 10.4) that requires the aperture of the 5" or larger telescope, not the distance between them. The fainter star will be quite difficult to make out from the very bright orange glare of the primary star. Look for "B" slightly north of due east from "A."

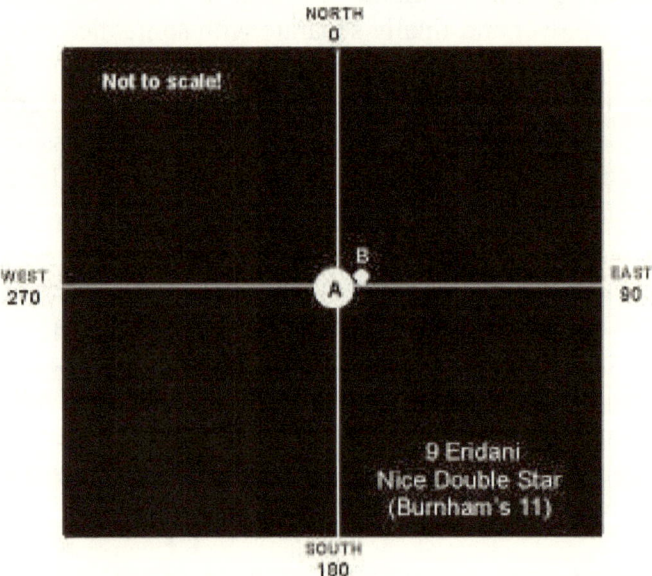

Object 7 - A Fun and Easy Semi-Regular Variable Star (accessible from northern latitudes, too!) – Z Eridani

This variable star is one of many interesting and yet-to-be-understood "Semi-Regular" stars; such stars demonstrate a "suspected or slight" regular period between brightening, dimming and then re-brightening again (about 80 days for this star), but are characterized during that period by sometimes major fluctuations....unexpected dimming or brightening....long periods at maximum or minimum....very rapid brightness increases followed perhaps by extremely slow fading....and so on. Hence the need for YOUR observations of such stars. Even with a period of only 80 days, this star should be observed at least once weekly when possible. For a complete discussion on this classification of variable and methods for observing variable stars, see my ASO Variable Star Observing Guide in the GUIDES of this website.

Z Eridani is one of the brightest of all semi-regular stars, fluctuating in brightness from magnitude 6.4 to 7.9. It is a VERY red, red giant star, of the extremely late "M" class spectrally. Its wild fluctuations are well known and the star is well worth the watching. Because of its brightness throughout its cycle, it is visible in small telescopes and even binoculars with the proper chart. A chart is necessary to: 1) find the right star to watch; and, 2) have stars provided nearby with magnitudes given that have been correctly checked as accurate through which brightness assessments can be made. For such brighter stars needing likewise bright comparison stars, one must choose a wide angle chart that features a large chunk of sky. The

American Association of Variable Star Observers (AAVSO) provides complete charts for virtually every variable observable with common telescopes. The wide-sky "finder" charts are designated "a" charts. The proper chart for Z Eri for both locating and observing its magnitude changes ican be obtained at:
https://www.aavso.org/apps/vsp/ . Note for these charts, simply type in the NAME of the variable at top to generate your choice of chart.

Object 8 - Finally - a "REAL" Deep Sky Object in Eridanus! - A Face-on Spiral Galaxy - NGC 1232

The striking photograph above of this classic galaxy was taken with the HAWK-1 telescope from the ESO. This is a fabulous face-on spiral galaxy located near the middle (north-to-south) of Eridanus at about -20 degrees from the celestial equator. This low angle for northern observers in not too much of a hindrance until attempting to observe this magnitude 10.4 object. It is a fairly large galaxy to view (7.7 x 5.5 minutes arc) so its brightness is spread out over that area. It is visible as a small round "glow" in a 3" scope under VERY dark skies with no moon; much clearer in a 5" and 6" scope and affords some interesting detail with the 8" and larger scopes. I have clearly seen three of the brightest "knots" of stars in the spiral arms and suspected a fourth visually with a 12" Schmidt. The many arms of this spiral are very clear, but extremely faint, visually with the 24" telescope at 212x.

Object 9 - Another Spiral Galaxy - NGC 1187
Almost a duplicate of the many-armed spiral ngc1232 (above), this spiral is considerably fainter. At magnitude 11.2, it is still visible in smaller telescopes, though void of any discernable detail in telescopes smaller than 12" diameter. In large scopes, the spiral arms very much resemble those seen (and photographed) with ngc1232. It is slightly smaller than the former and without the "clumping" in any arms; thus even larger telescopes visually show nothing but a circular disk of very faint light.

Object 10 - A Nice Planetary Nebula - NGC 1535
This is the ONLY planetary nebula visible visually in any telescope in the constellation of Eridanus! NGC 1535 is a fairly small (20" oval shaped) cloud of blue-appearing gas. It stands out unmistakable amidst the black sky and very star-poor background of north-central Eridanus. Look for this small disk

of light nearly due east of gamma Eri (ZAURAK) by nearly 5 degrees and since the object is high in the constellation, it is a good target for all telescopes in the northern hemisphere. The distinctive blue color is just beautiful and at first glance it stands out much like a small blue planet or other body, almost 3-dimensional. After you find ngc1535, increase the magnification to about 30x per inch to magnify the scale of this interesting sight. Right in the middle is a central star of magnitude 11.5 which can be glimpsed fleetingly with a 5" or 6" scope. It should hold steady with an 8" instrument. It is this star that "exploded" in the prehistoric past creating the planetary nebula that we now see. The nebula is a relatively bright 9.3 magnitude, so this nice object should be observed by all.

WANDERING ABOUT....YOUR NEW "USER OBJECT" IN ERIDANUS

Here is a beautiful object whether you are viewing from either the northern or southern hemisphere. OMICRON 2 ERIDANI (also known as Flamsteed's "40 Eridani") is comprised of two brighter stars, both about 4th magnitude. It is located very near "Omicron 1" (the northernmost star with a fairly bright star (37 Eridani) immediately to its west; both of these are clearly visible together in your finderscope or in binoculars); immediately below (south) is "Omicron 2", is our fantastic triple star. Although easily seen as double in even small telescopes, the duplicity of this star was not "discovered" until Sir William Herschel cataloged it late in 1783. A 2.4" refractor held steadily will reveal the "A" and "B" components of this wonderful star, separated by a whopping 83" arc, about two Jupiter diameters as seen in the same eyepiece. The only difficulty in observing with the smaller telescope is in the faintness of the "B' component, at magnitude 9.3, compared to the bright primary "A" star, at magnitude 4.4 ("Omicron 2"). See the chart below for locating the fainter star.

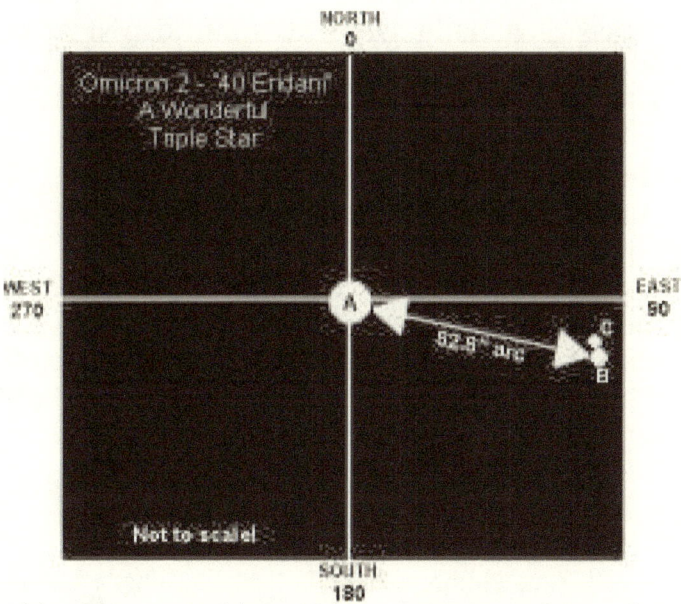

This is an exciting object, not just from the wonderful view afforded by the two "A" and "B" components, but the fact that the "B" star....fainter star....has yet ANOTHER star encircling it as well! We note in the chart above that "B" is in Position Angle 105 degrees, or nearly due west and a slight bit south of the bright star. It is this fainter star that is also double. Circling "B" is an 11th magnitude "C" star in Position Angle 347 degrees from "B" (we do not measure its angle from the primary, but from the star which it orbits which, in this case, is

"B"). So look for the third star just about straight north (a bit to the west) from the 9th magnitude "B" star.

Of course you can use the chart above to find the star quite simply with your finderscope, but it will always be ready for you to GO TO if we add it among the growing library of "User Objects" on your computerized Autostar. Coordinates (corrected for epoch 2001) for OMICRON 2 ERIDANI ARE:
R.A. 04h 14m
DEC -07 37

On your PC program (see HELP instructions) or AutoStar, go to: "Select/Object [enter]...." scroll down to "User Object" [enter]. Now enter the coordinates given above for OMICRON 2 using the number keys on AutoStar. After entering the coordinates and pressing "Enter" yet again, scroll down one and you can list the magnitude of the object as "4" [Enter].

*   *   *

LEPUS....."ehhhh, what's UP doc?" What's UP in the Southern hemisphere is DOWN for Northern observers.....Lepus, the celestial hare.....the "widdle, wascally wabbit of the winter wonder-world." The celestial bunny might have been better suited for the spring skies of Easter-time, but perhaps if he had delayed for the warmer spring weather he would be even further on the heels of the tortoise (which there is no such thing in the 88 constellations, but it makes for good conversation when you show your skywatching visitors to the winter star party the Leaping Rabbit below the feet of Orion! And...have

fun with the carrot by encouraging them to find it in
this small constellation....it's not there.

Look for our LEPUS *Constellation* tour in
Volume II of this series!

*   *   *

"Thousands of experts study overbought indicators,
oversold indicators,
head-and-shoulder patterns, put-call ratios, the Fed's
policy on money supply, foreign investment, the
movement of the constellations through the
heavens, and the moss on oak trees,
and they can't predict markets with any useful
consistency, any more than the gizzard squeezers
could tell the Roman emperors when the Huns
would attack...."

Peter Lynch

ABOUT VOLUME TWO

I have always been accused of talking too much. Someone asks me the time and I tell them why Rolex watches are too expensive.

Constellations: Volume II is testament to that fact. Originally intended to be a quite thick one-volume reference guide for computerized telescopes, I quickly realized that there is a lot that needs to be explained and told, both in terms of operation of modern GO TO telescopes, as well as with the thousands of beautiful objects awaiting the users of those telescopes.

Using a telescope is like taking a bath every day. Some people cannot wait for the experience, but others need prodding and might find excuses to put off the task.

Hopefully these guides of the Constellations for your computerized telescopes will take much of the "task" out of observing the heavens and inspire you to want to get out on every clear, moonless night throughout the year.

Volume II covers the constellations from Hercules through Virgo and includes some of the most exciting of all sky objects – those in Orion and Taurus and Ursa Major.

Remember that constellations visible for northern hemisphere observers are covered here (sorry, Down Under!), but the vast majority are indeed included. Enjoy!

Doc Clay

www.ingramcontent.com/pod-product-compliance
Lightning Source LLC
Chambersburg PA
CBHW031816170526
45157CB00001B/74